ENERGY CULTURE

Energy and Society
Brian Black, Series Editor

OTHER TITLES IN THE SERIES

On Petrocultures: Globalization, Culture, and Energy
Imre Szeman

Oil and Urbanization on the Pacific Coast: Ralph Bramel Lloyd and the Shaping of the Urban West
Michael R. Adamson

Oil and Nation: A History of Bolivia's Petroleum Sector
Stephen C. Cote

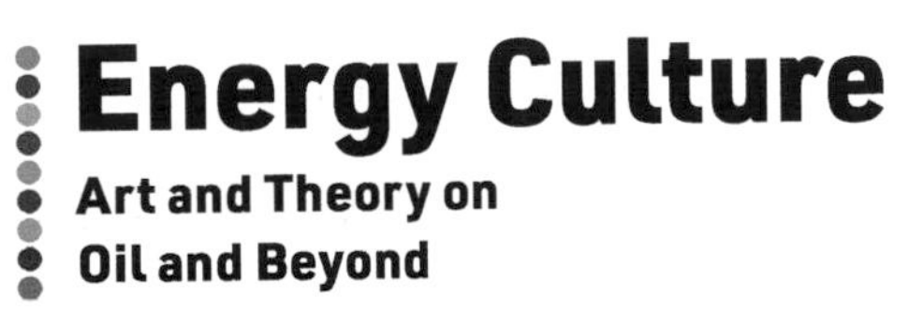

Energy Culture

Art and Theory on Oil and Beyond

Edited by
Imre Szeman
Jeff Diamanti

West Virginia University Press
Morgantown 2019

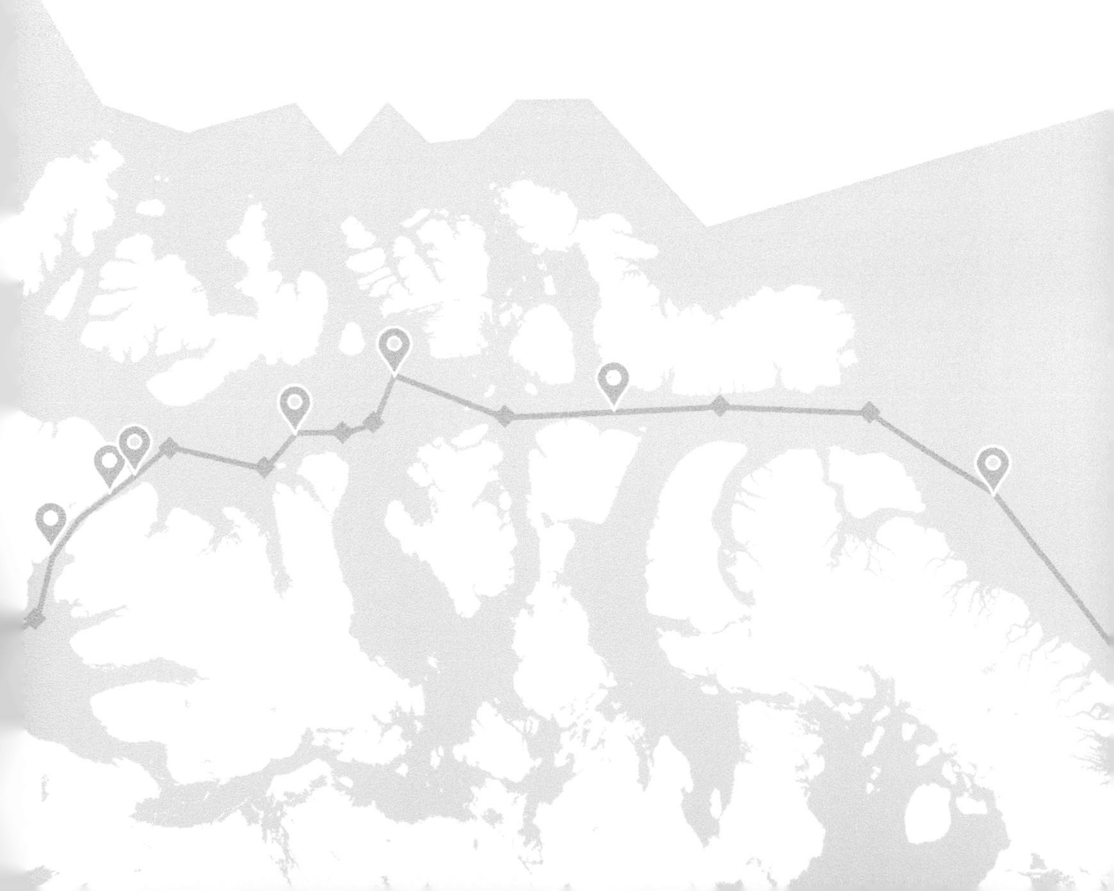

First edition published 2019 by West Virginia University Press
Printed in the United States of America

ISBN
Cloth 978-1-949199-11-6
Paper 978-1-949199-12-3

Library of Congress Cataloging-in-Publication Data
Names: Szeman, Imre, 1968– editor. | Diamanti, Jeff, editor.
Title: Energy culture : art and theory on oil and beyond / edited by Imre Szeman and Jeff Diamanti.
Description: Morgantown : West Virginia University Press, 2019. | Series: Energy and society series | Includes bibliographical references and index.
Identifiers: LCCN 2019019865 | ISBN 9781949199123 (pbk.) | ISBN 9781949199116 (cloth)
Subjects: LCSH: Fossil fuels–Social aspects. | Petroleum as fuel–Social aspects. | Renewable energy sources–Social aspects.
Classification: LCC TP318 .E545 2019 | DDC 621.042–dc23
LC record available at https://lccn.loc.gov/2019019865

Cover and book design by Than Saffel / WVU Press
Cover image by Than Saffel / WVU Press based on Fig. 2.4. Map taken from the 2013 Arctic Fibre Submarine Cable System Project Description / Project Proposal Plain Language Summary, 3, gcc.ca/pdf/2013-10-23-Non-Tech-Summary.pdf.

This project was supported by the Social Sciences and Humanities Research Council of Canada as well as the Canada Research Chair in Cultural Studies.

To A. P., for everything that has happened,
and everything still to come.

CONTENTS

Part III: The Politics of Energy Culture

ACKNOWLEDGMENTS

Collaborative projects such as this one incur far too many debts, favors, inspirations, and supports to list them all here. The initial art and research residency in Banff, Alberta, from which this book emerged involved more than twenty researchers and artists, substantial support from the staff at the Banff Centre for Arts and Creativity, and many hours of invisible labor both in our respective homes and in the kitchens and studios that supported our month of collaboration. At the Banff Centre, we would like to thank Debbie Morgan, Brandy Dahrouge, Kelly Moynihan, and Charlene Quantz-Wold for a perfect month in the mountains.

Sean O'Brien, Caleb Wellum, and Jordan Kinder were instrumental in preparing the manuscript. Special thanks to Derek Krisoff, our editor at West Virginia University Press, for helping us see this through, and Than Saffel for turning this book into something worthy of the aesthetic ambitions of its brilliant contributors.

Jeff Diamanti would like to thank Marija, Marcel, and Lorenzo for luxury bowling, sushi on toy trains, and all you can eat dessert(s).

Imre Szeman would like to thank the energy, passion, and commitments of friends and colleagues—constant reminders of why the work we do needs to be done.

INTRODUCTION

Imre Szeman and Jeff Diamanti

This collection of scholarly and artistic interventions is the outcome of a month-long residency, Banff Research in Culture (BRiC). Held annually at the Banff Centre for Arts and Creativity in Alberta, Canada, and organized each year around a different theme, in 2016 BRiC focused on "energy culture": the social habits and cultural imaginaries that have emerged out of the shared uses and abuses of energy around the world. A growing number of books now investigate the relationship between energy and culture, especially in the emergent field of "energy humanities."[1] What distinguishes *Energy Culture* is that it is the first work to cross-pollinate artistic and scholarly research, and to highlight investigations generated, workshopped, and produced in a collaborative environment with the intent to intervene in our deep, unquestioned, and still underexplored relationship to energy. This intervention matters because only by understanding the full extent of our commitments to and dependence on high levels of dirty energy will the planet's human inhabitants be able to make the social transitions and transformations needed to offset global warming.[2] Given the lack of action on climate change to date, no intervention could matter more.

Unlike conventional academic collections—which originate with an editorial call, followed by a series of individual submissions that roughly fit the theme, through to the end product, which is composed, revised, and finalized piece by piece—this book insists on the need for a different form of thinking, inquiry, and composition. The challenge of critically examining energy to conceive more fully its impact on culture and society—and thus, too, its overall significance for social and environmental transformation—demands experimental and speculative forms of research, modes of inquiry designed to enable the reconceptualization of energy as social form, and a broad reimagining of social form as a set of practices and frameworks always already entangled with energy systems.[3] By 1943, anthropologist Leslie White had already produced the outlines of a vocabulary to speak about (in the words of one of his influential essays) "Energy and the Evolution of Culture," while more recently, Dipesh Chakrabarty has argued that historians ought to reexamine the forms of freedoms that characterize modern culture through the lens of the ever expanding base of fossil fuel use upon which they are built.[4] The social sciences and humanities have long held that energy and culture interimplicate one another macrosocially and microsocially, developing implicitly and recursively over long periods of time. *Energy Culture* builds on this core insight by turning to critical and creative modes of inquiry both occasioned by, and designed to crack open, the invisibility and ubiquity of our twenty-first-century "petroculture," which poses a set of conceptual and political challenges not easily overcome by desires for sustainability, equity, or justice. The three sections that make up *Energy Culture*—"Mapping Energy Culture," "Figuring Energy Culture," and "The Politics of Energy Culture"—name the vectors of inquiry that its contributors jointly pursued while in Banff in their efforts to answer the project's overarching question: what needs to change in our research habits, interpretive frameworks, and social and political lives in order to more fully conceive of the planet's energy culture as it exists and as it might yet exist?

The contributions to *Energy Culture* represent different modes of examining, looking, and narrating, from academic inquiries to art projects, and from participatory social practices to models of new ways of living and acting. What connects them is their concerted interrogation of the assumptions and presumptions that animate our relationship to energy, and especially to our main source of energy for close to two centuries: fossil fuels. The primary issues that these varied contributions seek to address—the operations of energy in shaping contemporary culture, and our critical ability to grapple with this as we attempt to move away from fossil fuels—exceed the analytic capacity of any one discipline or conceptual approach. No master narrative on energy will set us on a cleaner, more equitable path; if anything, faith in master narratives is part of the problem rather than an element of the solution. The pieces in this volume draw attention to specific aspects of the form and character of the energy culture that shapes us. In doing so, they also point to limits in the approach to energy of other pieces collected here. Because we deliberately avoid a master concept of energy, the chapters gathered here work together and also (at times) work productively against one another. *Energy Culture* is a critical collection in the truest sense: a group of pieces that speak to a single issue, each providing a piece of a larger picture that could not be grasped without their combined intellectual contributions and insights.

Despite the diversity of approaches and focal points, the contributions to *Energy Culture* start from two common insights. The *first* is that energy is linked to culture in a *fundamental* way. The insight that culture is entangled with the infrastructure of energy, and, equally, that the continued expansion of the global energy system is contingent on global culture, moves these critical interrogations away from moralizing and shortsighted approaches to energy and environment. Rethinking energy and culture in terms of their mutual dependency builds on some of the most exciting new research in the environmental humanities, which emphasizes the depth of commitments made to high-energy lifestyles. In his stirring assessment of the multiple challenges posed by climate change, Amitav Ghosh reminds us that

> culture generates desires—for vehicles and appliances, for certain kinds of gardens and dwellings—that are among the principal drivers of the carbon economy. A speedy convertible excites us neither because of any love for metal and chrome, nor because of an abstract understanding of its engineering. It excites us because it evokes an image of a road arrowing through a pristine landscape; we think of freedom and the wind in our hair; we envision James Dean and Peter Fonda racing toward the horizon; we think also of Jack Kerouac and Vladimir Nabokov, of a quintessential narrative whose very setting is the road. When we see an advertisement that links a picture of a tropical island to the word paradise, the longings that are kindled in us have a chain of transmission that stretches back to Daniel Defoe and Jean Jacques Rousseau: the flight that will transport us to the island is merely an ember in that fire. When we see a green lawn that has been watered with desalinated water, in Abu Dhabi or Southern California or some other environment where people had once been content to spend their water thriftily in nurturing a single vine or shrub, we are looking at an expression of a yearning that may have been midwifed by the novels of Jane Austen.[5]

Ghosh places emphasis on the historical depth of these desires—the discourse and cultural logics of a winter vacation for suburbanites stretching back to Defoe and Rousseau, and the fantasies of lawns and gardens for homeowners across the planet linked to yearnings found in Jane Austen. In the desire for the middle-class safety of a backyard, a small slice of heaven that one can call one's own, or in the power and freedom of a car on the road, culture and energy constitute each other, and in the process constitute modern, global society as well.[6]

The economic and transport infrastructures that have been generated by the ever-expanding use of fossil fuels (and which in turn support and amplify that self-same use) constitute a significant dimension of global energy culture. Just as significant—perhaps even more so—are the cultural desires to which Ghosh and the contributors to this volume draw our attention. While not everyone on the planet has access to the same levels of energy, the desire for the objects and services that come with energy use animate global imaginaries of the good life. The link of these desires to energy, and our practices of energy to these desires, can be difficult to separate from the sheer phenomenological givenness of modern culture and experience. What possible kind of critical response can be mounted to the constitution of the whole of one's cultural desires and social experience? Can one pull them apart to isolate the role of energy in figuring them? Can one do anything other than point speculatively to sites and practices of connection and constitution, and imagine some other way of putting the pieces of the puzzle together that uses fewer kilojoules each year? Taken together, the pieces in *Energy Culture* speak to the challenges of taking energy seriously for culture—a task that goes beyond (say) looking at films or novels that feature energy (an important first step in the practice of energy humanities), and that demands an interrogation of desires for representation, narrative, and amusement, of the modern system of social and cultural capital and its links to energy, and of the divisions of the social, cultural, political, and economic into which modernity has been so comfortably separated, with energy usually out of sight and out of mind.

To take energy seriously in relation to culture demands the exploration of new forms of being, belonging, ethics, and politics—new, that is, to a modernity that has sought to impose the violence of its logic and practices on all human communities, and in the process has also redrawn the map of relations to the nonhuman (when it hasn't torn up and abandoned this map altogether). While fossil fuels have been both qualitatively and quantitatively dominant in modern industry and culture, they have also coexisted with both residual and emergent energy sources, each with its own geographies, histories, and social implications. Energy culture unfolds unevenly, which is why in Ghosh's account the forms of desire and imaginaries to which fossil fuels stick are not always determined by what has come to be known as "petroculture," but nevertheless help harden our global reliance on the freedoms that come with fossil fuels. But other forms of attachment coexist with the petrocultural, and these will have to matter to any environmental and social politics to come. One doesn't need to look far to find counterexamples of living with energy. Warren Cariou points out that "traditional Aboriginal energy-use practices are characterized by what might be called *energy intimacy*, in which every community member necessarily has direct and personal relationships with the sources of their energy." In

Cariou's stirring rejoinder to the modern practices of energy use, he goes on to say that "energy intimacy means that energy is always contextualized, always specific to a particular place with which the energy user must establish an intimately familiar connection."[7] The demand for and assertion of energy intimacy is a vibrant political act in an energy culture that is constituted (for most) around the distance of energy from source to the sites at which it is employed; notable exceptions are those minority communities, including Aboriginal communities, that have had to suffer the environmental and social consequences of the mass extraction and production of energy, and fossil fuels in particular.[8]

As another example of energy intimacy, consider the many rituals that rearticulate small groups to the ecological chain in a given activity, through what the German American philosopher Albert Borgmann calls focal things and practices.[9] These include stocking and managing wood-burning stoves, performing Japanese tea ceremonies, and cycling, among many others. In common across these varied activities is the way that the human body is made to feel its way through the material stuff that entangles our cultural practices and rituals. Key to the difference between focal practices and fossil-fueled ones are the mediations that intervene between the body and its experience of the world. Technologies that mask the contingences of energy use help to bury what Patricia Yaeger has termed our "energy unconscious" in a field of forces so remote from their crude materiality that our habitus begins to disavow them altogether.[10] Focal things, in contrast, forge an embodied epistemology so that the varied phenomenologies that come with the environmental contexts of different energy sources become rearticulated to our shared ways of knowing. One of the challenges faced in reimagining energy culture is that practices that put the subject in direct contact with the chain of materials involved, as in Borgmann, are difficult to imagine beyond small groups. Technologies and capacities bound to fossil fuels are ubiquitous and effervescent, well exceeding the sites usually targeted by environmentalist groups, such as the smokestack and the tailpipe. Mapping and decommissioning the ubiquity of fossil fuels, therefore, requires a variety of critical approaches, involving both exposition of how certain practices and beliefs involve an energy unconscious, and creative experimentation with new practices of energy intimacy that help to normalize the deintensification of fossil fuels.

If the first insight shared by the contributions to this volume is the deep imbrication of culture with energy, the *second* is that we presently occupy an interregnum in our relationship to the environment and the nonhuman. Through whatever medium they might employ, our sharpest critics have been tearing away at the infrastructures (physical as well as conceptual) of modernity—the separations, prohibitions, and exclusions they enable, the desires and fantasies they underwrite (tropical winter vacations for everyone! lawns for all!), and the material resources they demand. Elizabeth Povinelli reminds us, however, that at present our gestures to exceed the logics of modernity remain within the dynamics of what she has termed "settler late liberalism." Povinelli describes the current moment of the configuration of power as "geontopower." This is not a new form of power, but constitutes the last gasp of modern biopower. Geontopower "does not operate through the governance of life and the tactics of death," she writes, "but is rather a set of discourses, affects, and tactics used in late liberalism to maintain or shape the coming relationship of the distinction between Life and Nonlife."[11] Povinelli cautions that the diagnostics that we have developed in relation to our understanding of (and desire to undo or manage) the gap between Life and Nonlife remains "a collection of governing ghosts who exist in between two worlds in late settler liberalism—the world in which the dependent oppositions of life (bios) and death (thanatos) and of Life (bios) and Nonlife (geos, meteoros) are sensible and dramatic and the world in which these enclosures are no longer, or have never been, relevant, sensible, or practical."[12]

Posthumanisms, new materialisms, object-oriented ontologies, techno-utopianisms—these are all among the "governing ghosts" to which Povinelli draws our attention. The study of energy culture also sits within this conceptual and political gap named by geontopower. Unlike some of the other fields of inquiry named above, the demands energy places on critical thought produce a heightened awareness of the dimensions and dynamics of our current interregnum; the investigations in this collection are alert to the temptations of safe, well-known ideas and the need to decisively interrupt them. While not imagining that they leapfrog late settler liberalism into the coordinates of some new mode of power and belonging, the contributions to *Energy Culture* insist on the need to attend to the governing ghosts of energy that haunt being and belonging. When solar panels extend to the horizon, will we strip away the lawns on our front and backyards? When fossil fuels become expensive and in short supply, will social life become a struggle over who gets to vacation in Oaxaca? Or will these infrastructures and desires no longer be part of what it means to occupy this time and space, with others, living and nonliving?

There are historical reasons why this interregnum is troubled in provocative ways when the focus is on energy culture. For instance, the specter of energy scarcity meaningful to environmentalisms of an earlier moment appears today flipped, turned by new extraction technologies into the specter of abundance, glut, and excess. In contrast to narratives of peak oil, there has never been as much oil ready for market as there is today, and with economic and demographic growth plotted to continue deep into the century, this oil will have no shortage of buyers. Natural gas and coal have paradoxically become more abundant and profitable in the years following Chinese modernization and US–led industrial agriculture. A social or environmental politics that waits for the natural end of supply will have decades, if not centuries, of misery ahead. Late-capitalist liberalism appears poised to trundle merrily along, at least with respect to the energy that fuels it; by now we have numerous accounts, including that of Andreas Malm, that effectively debunk the fiction that energy transitions occur historically because of the hidden and rational hand of the market.[13] We are witnessing today a peculiar inversion of social and natural history, where what was the environmental background to history now appears in flux and revolt, while human history appears fixed, becoming the new ground to the figure of the environment.[14]

If energy is capacity, it is also a constraint—and in anthropogenic climate change, driven by centuries of practices animated by fossil fuels, energy has become the core constraint of our time, making it profoundly difficult to generate new cultural and social imaginaries, and original political and economic forms. What then are social scientists, humanists, and artistic researchers to do in this context of capacity and constraint, of the weight and apparent fixity of energy culture, which we understand needs to be changed even as we begin to grasp just how difficult it will be to change it—since it means to change, well, everything?

The contributions to *Energy Culture* aim to unsettle status-quo understandings and expectations of our relation to energy by collaboratively and creatively investigating the forms and figures that dent energy culture, both historically and in the present, in an effort to conceive of time-released tensions in future energy culture. The BRiC residency asked artists and researchers to collectively address energy's historical figures and futures, its visual and social economy, and its capacity to disfigure. In addition to taking up the challenge of critical analyses in the landscape of petrocultures, the contributions to *Energy Culture* collectively offer an affective, creative, and critical start to the transition and transformation of petrocultures into new cultures alive to environmental and social justice, traversing the multiple lines of force that energy has woven into the heart of contemporary society.

Energy Culture is organized according to the three primary vectors of inquiry that emerged from our collaborative research. In Part I, "Mapping Energy

Culture," contributors detail the site-specific forms of energy that in the North American context make daily life contiguous with fossil fuels. Part II, "Figuring Energy Culture," turns to the artistic and creative research occasioned by the ubiquity of fossil fuels in contemporary life, and the interdisciplinary challenge of creating new energy cultures. Part III, "The Politics of Energy Culture," concludes the book by scanning the socioeconomic contradictions, impasses, and imperatives generated by the long transition into our fossil-fueled present. The work of the artists included here is intended to trouble and unnerve long-established ways of thinking about and relating to energy. The critics and theorists who participated in BRiC in 2016 and whose work is found in these pages engage in a similar labor, if from distinct angles and along different conceptual vectors. It is too simple a division to suggest that the artists attend to the visual vocabulary of energy cultures, while critics interrogate the language of energy; nevertheless, some of the contributors to this volume are attuned to the circulation of concepts related to energy through language—to the limits as well as possibilities contained in the explanations we commonly offer of the energy landscapes we inhabit.

Part I: Mapping Energy Culture

The Austrian artist Ernst Logar has engaged in an extended assessment of the social, cultural, and political significance of resource extraction. His best-known project is *Invisible Oil* (2008), which documents his careful and nuanced exploration of oil culture in Aberdeen, Scotland. The images included here are from "Tar Sands: Approaching an Anthropocentric Site," an exhibit of his work held at Paved Arts in Saskatoon, Saskatchewan. At the core of the exhibit are screenprints of iconic images associated with oil in Fort McMurray, Alberta: a Suncor extraction site; the Athabasca River; a reclamation site (part of the network of reclamation projects described by Jordan Kinder in his contribution to this book); and a pipeline. These screenprints are generated using bitumen—the very stuff of the oil sands. They make up the rear walls of aquariums filled with water taken from the Athabasca. These pieces insist that we see water and oil as bound up with one another in ways that should give us pause. The stuff of life in the aquariums is toxic, just as the Athabasca River has proven to be for First Nations communities that live downstream from the oil sands. Written into the priorities we assign to our use of oil and water is a narrative of power and privilege common to every practice of resource extraction.

Mél Hogan's "Trespassage" explores the recent opening up of the Northwest Passage, the storied route linking the Atlantic and Pacific Oceans via the

Arctic Ocean. The Northwest Passage has claimed the lives of many explorers, including (most famously) Sir John Franklin, whose steam-powered ships set off to Canada in 1845. Since 2007, it has become possible to navigate a passage free from ice—a climate change–generated transition in how we understand the Arctic, which has now become a feasible site for extraction and the generation of profit. This has been as true with the passage of goods as it has been to the passage of information. "Each new attempt to traverse the Arctic speaks to a persistent urge to pass through more quickly than before," Hogan writes, "to find a shortcut—to save time and money—a knowing *trespassage* that justifies itself with an efficiency of its own measure." Hogan proposes *trespassage* as a concept through which to "rethink the conceptual entanglements of nature and media, where both serve as infrastructure that links time and space." The medium Hogan explores is an undersea fiber-optic cable, which the company Quintillion Subsea Holdings intends to lay in the newly ice-free passage. While the company has promised to connect the inhabitants of northern Canada to the rest of the world, its true intent is to use the Northwest Passage to connect Tokyo, Montreal, New York, and London. As Arctic ice disappears, the high use of energy at these nodes of capital has created the perverse possibility of linking them all at even higher speeds, in the process bypassing the inhabitants of those spaces whose lives have been forever altered by the ferocious commerce of Western energy culture.

Metaphors can enhance our ability to grasp complex realities and draw together, in a single figure, practices and processes that we might not otherwise understand as related. But they can also obscure reality, acting as convenient and confident narrative shorthand for realities that don't actually exist. So it is with the figure of the "cloud," that amorphous, transparent, immaterial space that houses our data and enables the near instantaneous transfer of information. As Jayne Wilkinson shows, one of the many things hidden by our commitment to cloud thinking is the material reality of our information age. The metaphor of clouds makes it seem as if we communicate through the atmosphere via invisible, immaterial networks—a clean way of connecting with one another that leaves no trace or physical impact. In reality, the global infrastructure of clouds can be found in infrastructure housed in our oceans. In addition to her own reflections on the social and political significance of the metaphor of clouds and the way that it impedes and obscures our view of today's oceans, Wilkinson examines how our extant tropes of reading clouds and traversing oceans operate in relation to discussions of information, climate change, and energy. Her chapter draws on the work of writers such as Italo Calvino and Rebecca Solnit, and others including Roni Horn and Nicole Starosielski.

Common to all of the artist projects in this book is the attempt to remap the spaces and places of energy—those networks and infrastructural systems of modernity that have reinforced our practices and habits in relation to energy. In her contribution, M. E. Luka describes some of the projects of the artist group Narratives in Space + Time Society (NiS+TS). Based in Halifax, Nova Scotia, NiS+TS plots sensorial walks for members of the community, which are intended to bring to life both well-known and lesser-known histories of the spaces they traverse. These walks are animated by Guy Debord's theory of *dérive*—a means by which to interrupt sedimented patterns of traversing space and the fixed experiences that accompany such movements. Luka describes the critical work undertaken by NiS+TS through the walks they curate examining the Halifax Explosion, an event that took place in World War I and damaged much of the city, resulting in widespread death and injury. "By walking through these spaces with community members, we help make visible not just how the Halifax Explosion's reverberations continue to be felt today but also how we could imagine a different present and future," she writes. Walking—using the energy of one's body instead of moving through city streets with the speed of the automobile—allows one to experience the space one inhabits very differently, and also leads to the production of a different kind of community than the one anchored to car culture. Above and beyond the lessons about the city learned by participants in these walks, their reduced time and scale generate opportunities for collectivity that seem to have drained away in our auto-mobilized cultures.

In their introduction to *Energy Humanities*, Imre Szeman and Dominic Boyer write: "There is a place for sober criticism and discussion in the enterprise of energy humanities; there is also a place for surreal vision and wild imagination."[15] "Several Documents Pertaining to the Cascade Energy (transition) Park Corporation Corporation (CORPCORP)," created by Marissa Benedict, Cameron Hu, Christopher Malcolm, and David Rueter, is a marvelous example of how imagination can open up new vistas onto energy modernity. We are given little context for the images, maps, and texts that make up this contribution, other than a short note on the first page, which tells us that the documents we are about to flip through were in a thumb drive found in a parking lot in Joshua Tree, California. The documents throw us into the middle of a mystery. Just what are these scanned documents and audio transcripts about? Who generated this information, and why? What exactly is taking place in the Cascade Energy Park? Are these documents evidence of malfeasance, or are they simply misplaced documents into which we are mistakenly reading a narrative? Combining the creepy, unsettling affect of Jeff Vandermeer's *Southern Reach Trilogy* (is the area described here a variant of his Area X?) and Reza

Negarastani's experimental fiction *Cyclonopedia*,[16] this piece offers no answers to these questions. And perhaps this is the point: by highlighting the narratives we normally depend on to relate energy to the environment, this piece unsettles our comfort with them, insisting that we need to connect the dots in a different manner than we have done up to now.

As the contributions by Hogan and Wilkinson highlight, the repurposing of nature to the ends of profit is an all-too-common feature of our energy culture. Indeed, even the process of cleaning up sites of oil extraction can repeat and reinforce our dominant relationship to energy, and can do so in a manner that legitimates the ongoing viability of dirty, damaging extraction. In "Sustaining Petrocultures," Jordan Kinder interrogates the logic of land-reclamation projects carried out by oil sands companies on those zones already stripped of bitumen. For oil sands companies, even if the process itself is complicated, the rationale for reclamation is simple: it aims to undo the damage companies have wrought in the process of strip-mining for oil, for both ethical and aesthetic reasons, and to sustain public trust in what they do. Reclamation projects like Gateway Hill, near Fort McMurray, are intended to return extraction sites to their original, natural state. Kinder offers a ferocious challenge to the practice of reclamation, which he sees as little more than a process intended to maintain, expand, and reproduce our energy culture. Reclamation projects produce not a renewed natural landscape, but a deeply artificial one. The artifice is a result not only of the poor job done in restoring these sites, but of the too-easy understanding of "natural" that would make it imaginable to technologically negate the human alteration of a landscape by altering it even further.

Part II: Figuring Energy Culture

The artists Heather Ackroyd and Dan Harvey (known as Ackroyd & Harvey) have repeatedly drawn attention to the complex coordinates of energy culture. In "Capitalism in the Corpse of a Whale," they remind us of the energy source and culture that existed just before the oil era. Whale oil came before fossil fuels: it lit up cities that burned the oil from the head of sperm whales and filled the pockets of whaling fleet owners and investors. Whale harvesting during the eighteenth and nineteenth centuries emptied out the oceans of whales; many species of whales, which continue to be hunted by some nations today, remain threatened with extinction. Ackroyd & Harvey detail an artwork they produced from the skeleton of a stranded minke whale, designed to bring attention to the ongoing human assault on the ocean (whether through fishing and whaling or through the dumping of plastics), and to alert us to the consequences on the planet of protracted profiteering. Dipped in a highly saturated alum solution, the whale bones in their piece emerge crusted in bright crystals—an apt and powerful metaphor of the effects and implications of capitalism, whether of the whale era or of the oil age that has followed in its wake.

One of the most vibrant areas of the energy humanities has been literary studies, a field that has seen critics revisit literary history and literary texts to look for insight into our energy culture. In his elegant and erudite contribution, David Thomas asks himself whether literature and literary criticism can truly contribute to a renewed understanding of the social and cultural significance of energy in the way that some critics hope it might. The reason for his doubt is simple. Following writers such as Amitav Ghosh, Thomas explores whether the ecocritical attributes of contemporary literature have been oversold, especially with respect to encounters with our petrocultures. His chapter offers a relentlessly critical account of the limits of literature in relation to the environment, and points to the problems of deep or close reading of literary texts as a way to learn about energy. "Study of literary writing in the age of the Great Acceleration," Thomas writes, "seems to offer more insight into the dreams and fantasies that permeated elite culture as the signature modalities of petrocapitalism developed, than into the underlying nature of the project itself."[17] What Thomas takes away from his assessment of the current practices of the energy humanities in relation to literature is an insight that underlies all of the contributions to *Energy Culture*. Even given his criticisms and hesitations, energy remains for Thomas "a lodestar for the kind of synthetic vision that we would need if we were to produce forms of social life that were capable of submitting the power of techno-science to a kinder and more holistic understanding of our place in the world."

The brief interview with the artist Maria Michails gives us insights into her own unique exploration of energy culture. Michails's interest in energy dates back to *EMERGY* (2008), a project that uses the human power of its participants to draw attention to our culture of convenience and wealth. Her projects require those experiencing them to exert energy—for instance, by pedaling a stationary bicycle—in order to bring her pieces to life. This constitutes a form of embodied knowledge and education essential to the task she intends her projects to carry out. From *S.OIL* (2012) to *Mapping the Terrain* (2016), Michails transforms the sometimes abstract and distant systems of energy into material processes that one can experience physically as well as intellectually. This makes it possible to imagine energy as a dimension of modern life into which communities can politically intervene and shape to different purposes than our extant infrastructures.

For a significant part of philosophy's history, a nascent concept of energy lurked behind the development of ontologies and the outline of metaphysics. Energy played a crucial link, for instance, in the works of Leibniz, Helmholtz, and Nietzsche, between materiality and cosmic interconnectivity. Am Johal's "The Energy Apparatus" asks us to consider why energy has become conceptually and philosophically defanged, shifting from the expanse of philosophy to a relatively limited sphere of political contestation. Missing, Johal argues, is a theoretical lens that allows us to see both the macrosocial and historically specific ways in which the material origin of energy is tied to its true political force. Turning to Italian philosopher Giorgio Agamben's treatment of power and sovereignty over different forms of life, Johal provocatively considers whether oil has become the nonhuman sovereignty on which the legal apparatus of modernity is built. In productively asking how Agamben's discourse allows us to see energy anew as a social substance, Johal also brings insights from the energy humanities to bear on Agamben's own philosophical project, probing the limits and presuppositions that emerge from a blindness to energy's full significance and import.

One of the reasons that energy has long been invisible is the distance of oil fields from the urban conglomerations in which most of us live. Artists Hannah Imlach and Thomas Butler confront this distance by bringing energy to the city—not as potential threat, but as promise. *Aeolian Survey* is a planned art installation that would involve the installation of aeolian harps throughout Glasgow. The sound of these hidden harps—played by the wind moving across the rooftops of buildings—would be transmitted to an art gallery, where the accumulated sounds would be played over speakers. The main point of *Aeolian Survey* is to draw attention to the renewable energy potential of a city center. Scattered all over the city, the harps would "act as a synaesthetic mapping device, conveying the potential of local wind power through sound alone." But this piece would do more than just emphasize the potential of Scotland to generate a larger degree of its energy via renewables. "Mindful of the present," write Imlach and Butler, "we mix the roar of the electrified harp with location recordings made inside the very buildings they crown"—a sonic cross-section that collates evidence of electronically enhanced human activity. The gallery sounds bring to our attention both the generation of energy and its expenditure, creating an affective and personal encounter with a large, abstract system that can be difficult to confront or grasp.

In "Anecdotal Encounters on Driveways," artist Megan Green introduces us to the themes that have shaped her work on oil cultures. Feral suburbs, kitsch, and Newfoundland identity: these topics sound like stereotypes of life in Fort McMurray, the city at the center of Canadian oil sands extraction. A Newfoundlander now living in Fort Mac, Green explores a very different petro-aesthetic than the one to which we have become accustomed. Instead of exploring its inhuman colors and otherworldly scale (as in the photographs of Edward Burtynsky or filmmaker Peter Mettler's *Petropolis*),[18] Green turns to the quotidian character of life in the oil sands. For those living in Fort Mac, the oil sands are just part of the neighborhood—part of the sight, feel, and smell of the place. Those living near sites of extraction cannot rely on an aesthetic that criticizes from afar; in any case, as Green points out, this aesthetic repeats the practice of distantiation through which oil extraction has long enabled and legitimated its practices. Green also critiques the turn to nature as a solution or resolution of the environmental problems generated by energy extraction. Her art comprises the remnants of another form of resource extraction: game hunting, which for many who work in oil fields constitutes a turn to the natural and away from the damage enacted by the culture of extraction in which they spend their daily lives. Green's pieces and the anecdotal encounters that generate them point to the need for a new vocabulary of ethics and aesthetics through which we might reimagine our current relationship to resources.

Like the other artists in *Energy Culture*, Jacquelene Drinkall's work explores the sites, spaces, and modes through which our culture is imbricated with energy. Through her interrogation of the degree to which we are, in fact, creatures of petrocultures, she introduces a unique perspective on energy culture. Drawing on philosopher Catherine Malabou's concept of "plasticity," Drinkall insists on the impact of climate change on the cognitive capacities of the human brain—not just the configuration of our concepts, thoughts, or imaginaries, but the physical shape and wiring of the nervous system. "Lakes and mountains are not foreign to the ecosystem of the human brain," Drinkall writes, "and if humans destroy them, they destroy their brains and themselves." This strong equation of nature and neurology informs the two projects that she describes in her chapter. Both projects highlight the complex ways in which, in her words, "physical and neurological structures are bound to climate, energy cultures, and cybernetic systems."

Part III: The Politics of Energy Culture

One of the challenges that the planet will have to undergo over this century is to shift from fossil fuels to renewable forms of energy, and to do so while also expanding access to energy for most people on the planet. Exactly how we might expand energy use for citizens of developing countries while minimizing or even reducing CO_2 levels is hard to fathom. Even the shift to renewable energy has proven difficult: the infrastructures and political systems generated by dirty energy have had a tendency to impede movements toward clean energy. Jenni Matchett, a former employee of one of North America's major solar energy companies, describes the models for solar deployment that have emerged in North America to date. The growth in solar energy technologies and the expanded consumer demand for solar does not on its own guarantee that individuals, communities, or municipalities can easily switch off coal or diesel generators and begin soaking in the energy of the sun. The electricity grid is an old system that needs updates to bring solar online, and the policies of governments and utilities are weighed down by accumulated practice, histories, and expectations. Matchett describes three models that have emerged for solar energy: solar energy marketed as a consumer product, solar energy as a retail investment asset (that is, an investment mechanism through which the development of solar is financed), and community solar. Matchett argues that only community solar will allow solar energy to become more than another commodity fashioned along capitalist market principles. In Matchett's contribution, we begin to see the ways in which the optimism attached to energy transition is already framed by a way of "just doing business" that emerged alongside the development of oil capital.

If solar power does not in and of itself move us out of oil capital, might there be other changes to existing infrastructure and social practices that would enable true energy transition to take place? Keller Easterling maps out the possibilities that would be opened up by the introduction of "the switch"—a space that allows for upshifting and downshifting into different forms and modes of transportation. Automated vehicles (AVs) have been presented recently as a mechanism through which it might be possible to address the worst excesses of car culture. However, without redefinitions of the space in which they are employed, AVs may generate as many problems as they create (for example, current mass transit riders may shift *en masse* to the greater convenience of AVs). Easterling shows how the creation of transportation hubs that would bring together hitherto distinct activities, such as shopping, education, and exercise, would ensure an optimal employment of travel—from mass transit to the individual use of AVs. Easterling points out how the switch might generate a reconstitution of the "essential disposition of power in cities" by creating possibilities that anyone could travel wherever they might like. She convincingly advocates the promotion of the switch as a form of politics by other means—one that generates a redefinition of the current configuration of urban and suburban space, as well as the desires and practices densely coded into our current way of living together.

Darin Barney's focus on the infrastructures of fossil capital redefines pipelines, transistors, and refineries as sites not only of production, but also of class *re*production. This redefinition matters enormously for how we conceive of a politics adequate to desires for a large-scale energy transition. Barney forwards a series of theses on the relationship between energy infrastructures and the practice of sabotage in order to resuscitate what was a key mode of left politics in the past, as well as to help us understand how fossil fuels lubricate capitalist accumulation through a kind of preemptive sabotage of its own. Barney shows why intervening where the flows of energy mix and mingle with and as the flows of capital means exposing the smooth space of the economy as coextensive with the crude space of fossil fuels. Leaning on the essential technologies where energy is most tangible leads to a labor consciousness alert to its own energy *un*conscious.

Antonio Negri also draws into focus the tactics of a radical tradition born in the moment of fossil-fueled industrialization in order to identify the shape of energy resistance to come. Using the imagery of the furnace, Negri turns the chronological sequence of sabotage-strike into an index of the spark and fire of fossil capital, both as political struggle and literal, energic force: "Sabotage is the spark of this protest. Often individual. But sabotage can also be collective (perhaps it's always collective), because it's hard to imagine a spark without a fire that burns. In fact, the strike can arise, as a collective behavior, when the spark, through organization and the awareness of the organization of labor, succeeds in setting fire to the prairie." If the prairie is the topography of contemporary capitalism, then the spark is the collective decommissioning of energy as a form of social domination. To set today's economic fabric on fire is to break capital's facility with fossil fuels. At the same time, economic and political developments recursive to today's energy system need to be politicized, including governance via biopolitics and the informatization of society. To carry out these political projects, Negri's chapter makes the case for a new kind of strike, sparked by a new kind of sabotage.

Our energy culture is deeply problematic and deeply troubling; the vocabulary and practices of energy companies have long been deployed to render the practices of extraction normal and natural. Matt Huber's *Lifeblood: Oil, Freedom, and the Forces of Capital* (2013) was one of the first books that looked closely at the complex ways in which contemporary social life was a product of the energy culture generated by capitalism. In his contribution, Huber argues that it is essential that we see energy as a material force of social division and antagonism, especially between classes within society. Scholars have already explored energy as a site of social struggle and contestation; the originality of Huber's essay is his focus on energy consumption, as opposed to energy production, as a space of class division. At the heart of this essay is an exploration of the emergence of oil- and electricity-powered suburbs as a social space that creates racial, gender, and class exclusions and division. Huber makes the powerful argument that today we do not use too much energy, but *too little*; for most people in the world; certainly, for lower- and working-class communities and for racialized populations, energy use has always been circumscribed and limited. "Fossil fuels need not only be seen as the original sin of industrialism that we must reject and power down from," Huber writes, "but a dirty springboard to an abundant and clean energy future."

Mirroring the ubiquity of energy impact on our daily lives is the everywhereness of its environmental traces, from the plastic buildup in our oceans to the thick swirls of CO_2 floating in the atmosphere. Frequently overlooked in discourse on environmental justice is the slow violence enacted on vulnerable bodies by way of the metabolic cycles that connect human bodies to ocean and atmosphere. Concluding *Energy Culture* is a poem that bisects the lines of human and nonhuman, ocean and atmosphere. In her "Vortex of Light," Maya Weeks updates fundamental insights from the ecofeminist tradition of Carolyn Merchant, Vandana Shiva, and Maria Mies by connecting the petrocultural to what her poem calls "the chemical level." Written in a combination of poetic styles, Weeks's work moves across the sites and bodies that accumulate plastic, pausing on conceptual islands to ask such things as "BUT WHAT DO WE LOVE" before moving on to develop a speaking, lowercase "i" both drawn to and repelled by the materiality of plastic. The liberal standpoint animating so much environmentalism is made insufficient in "Vortex of Light," given the scope and scale of our petrocultural entanglement with natural systems: "Even if you live a good life in relation to pollution," the speaker suggests, "there are things you cannot see. A single / washing of a fleece jacket can release thousands of tiny plastic particles. More than one hundred / million microplastic particles are released into Advent Fjord every day."

As *Energy Culture* makes clear, the ongoing project of petrocultural critique is not only to unmask the world as one riddled with addiction to fossil fuels, but also to carve out moments and enclaves in which to refigure the project of our energy futures. This crucial tense—the future—that weighs on the present like a nightmare is, crucially, plural, since so many attempts are made daily to prefigure and overdetermine the future of energy, in what will always be a deeply political interpretation of sociocultural trends and tendencies (consider the gap separating those who imagine that solar energy leads to "solar communism" and those who see alternative energies as an exciting new sector in which to generate massive profits).[19] These trends and tendencies are contested and cumulative and so consist of myriad contradictions that are proving to be exceptionally hard to pry apart or even to grapple with. Critiquing the base assumptions upon which such projections into the future are made is an indispensable tool that humanists and social scientists bring to the problem of energy. But critique subtracts by design, leaving the plurality of the future tense opaque and unfocalized. The *creative* work of grappling with the conceptual, material, and historical character of energy—the work in which we see this collection participating—helps to focalize the embodied social practices of an asymmetrical energy culture, providing a model of what the most vigorous and alert forms of multimodal critique can add to the task of shaping new futures for our energy culture.

Notes

1. For an overview of the field of energy humanities, see Imre Szeman and Dominic Boyer, eds., *Energy Humanities: An Anthology* (Baltimore: Johns Hopkins University Press, 2017); and Graeme Macdonald and Janet Stewart, eds., *Routledge Handbook of Energy Humanities* (London: Routledge, forthcoming 2020). A text that promises to take the field in exciting and intriguing new directions is Brent Ryan Bellamy and Jeff Diamanti's *Materialism and the Critique of Energy* (Chicago: MCM Prime, 2018), a collection that includes contributions from George Caffentzis, Andreas Malm, and Alberto Toscano, among others.
2. If it wasn't already abundantly clear, change has to happen *now*. In "Revolution in a Warming World," Andreas Malm points out the degree to which the use of our era's dominant energy source—fossil fuels—and global warming are linked. He writes that "using conservative figures, excluding any future discoveries and deposits [of fossil fuels] made available by new technologies, Katarzyna Takorska and her colleagues place the effect in the ballpark of 8°C—hitting 17°C in the Arctic—rather than the previously believed 5°C. Converted into actual conditions for life on Earth, those average eight degrees would, of course, spell the end of all stories" (Andreas Malm, "Revolution in a Warming World," *Socialist Register* 53 [2017]: 120–42, 132 [quote]).
3. This way of framing the question of energy deliberately builds on insights developed by Donna Haraway and others in the environmental humanities, captured in the dehyphenated concept of "natureculture." The term "energy cultures" as a collection and as a concept turns on Haraway's gesture—namely, to incorporate always the mutually supportive structure of being shared across the human and nonhuman divide in ethics and politics. Importantly, it also underscores the extent to which energy has *not* been thought actively as social form in the environmental

humanities to date, and offers a sense of what a corrective to this oversight might look like.

4. Leslie White, "Energy and the Evolution of Culture," *American Anthropologist* 45, no. 3 (1943): 335–56; and Dipesh Charkrabarty, "The Climate of History: Four Theses," *Critical Inquiry* 35 (2009): 197–222.
5. Amitav Ghosh, *The Great Derangement* (Chicago: University of Chicago Press, 2016), 9–10.
6. On the expanding global fantasy of grass lawns and its environmental implications, see Yuval Noah Harari, *Homo Deus: A Brief History of Tomorrow* (New York: Penguin Random House, 2015), 58–64.
7. Warren Cariou, "Aboriginal," in *Fueling Culture: 101 Words for Energy and Environment*, edited by Imre Szeman, Jennifer Wenzel, and Patricia Yaeger (New York: Fordham University Press, 2017), 18–19.
8. On pipelines as mechanisms designed to inhibit energy intimacy, see Timothy Mitchell, *Carbon Democracy: Political Power in the Age of Oil* (New York: Verso, 2011), 36–38; and Christopher Jones, *Routes of Power: Energy and Modern America* (Cambridge, MA: Harvard University Press, 2014), 124–43.
9. Albert Borgmann, *Technology and the Character of Contemporary Life* (Chicago: University of Chicago Press, 1984).
10. Patricia Yaeger, "Editor's Column: Literature in the Ages of Wood, Tallow, Coal, Whale, Oil, Gasoline, Atomic Power, and Other Energy Sources," *PMLA* 126, no. 2 (2011): 305–26.
11. Elizabeth Povinelli, *Geontologies: Requiem to Late Liberalism* (Durham, NC: Duke University Press, 2016), 4.
12. Povinelli, *Geontologies*, 16.
13. Andreas Malm, *Fossil Capital: The Rise of Steam Power and the Roots of Global Warming* (New York: Verso, 2016).
14. See Bruno Latour, "Agency at the Time of the Anthropocene," *New Literary History* 45, no.1 (2014): 1–18; and Latour's interpretation of Jameson's now infamous claim that it is easier to imagine the end of the world than the end of capitalism in "On Some of the Affects of Capitalism," lecture at the Royal Academy, Copenhagen, February 26, 2014, www.bruno-latour.fr/sites/default/files/136-AFFECTS-OF-K-COPENHAGUE.pdf. See also Fredric Jameson, "Future City," *New Left Review* 21 (2003), where he writes: "Someone once said that it is easier to imagine the end of the world than to imagine the end of capitalism. We can now revise that and witness the attempt to imagine capitalism by way of imagining the end of the world" (76).
15. Szeman and Boyer, "Introduction," in *Energy Humanities: An Anthology*, 10.
16. Jeff Vandermeer, *Annihilation* (New York: Harper Collins, 2014); Reza Negarastani, *Cyclonopedia* (Melbourne: Re.Press, 2008).
17. For an overview of the "Great Acceleration," see J. R. McNeill and Peter Engelke, *The Great Acceleration: An Environmental History of the Anthropocene since 1945* (Cambridge, MA: Belknap Press/Harvard University Press, 2016).
18. Edward Burtynsky, *Oil* (Göttingen: Steidl, 2011); and Peter Mettler, dir., *Petropolis: Aerial Perspectives on the Alberta Tar Sands* (DVD, Greenpeace Canada, 2010).
19. See David Schwarzman, "Beyond Eco-Catastrophism: The Conditions for Solar Communism," *Socialist Register* 53 (2017): 143–60; and McKenzie Funk, *Windfall: The Booming Business of Global Warming* (New York: Penguin, 2014).

PART I

Mapping Energy Culture

CHAPTER 1

Oil on Water

Ernst Logar

Is there any sense in which, when it seeps up through the ground in the La Brea Tar Pits in Los Angeles, or oozes, untouched, in the Canadian wild, bitumen could remotely be thought of as hope?

—Karen Pinkus, *Fuel: A Speculative Dictionary*

Ernst Logar's *Oil on Water*, a series of water tanks screenprinted with bitumen, is the outcome of research trips taken by the artist to the Athabasca River region close to Fort McMurray, Alberta, from 2015 to 2016.

Fig. 1.1. *Suncor Mining Site—Fort McMurray* 2015, Ernst Logar, 2016. Bitumen screenprint on glass, 260 × 360 mm.

This is not the artist's first confrontation with the sites and spaces of resource extraction. Logar's *Invisible Oil* (2008), an interconnected series of multimedia pieces, explored the impact of fossil fuels on life in Aberdeen, Scotland, a community that has experienced the destabilizing effects of proximity to the epicenter of a major oil field. One of the most memorable elements of *Invisible Oil* was a series of photographs of scale models of North Sea oil platforms constructed by the artist—mock mechanisms of extraction crafted out of driftwood and detritus collected from North Sea beaches (plastic bags, bottles, jerry cans, and other containers; work helmets, shoes, and boots; balls used in various sports; foam and chunks of unidentifiable construction material).

Oil on Water also makes use of materials that Logar found on site during his research trips. The screenprints on glass at the heart of this exhibit employ bitumen—the "oily" substance of the oil sands that has generated so much social trauma and dislocation. As in *Invisible Oil*, Logar remains committed here to exploring the physical sites of resource extraction.

The screenprints on the following pages are representative images of key locations of the Canadian tar-sands industry. What distinguishes this

Fig. 1.2. *Athabasca River–Fort MacKay* 2015, Ernst Logar, 2016. Bitumen screenprint on glass, 260 × 360 mm.

encounter with oil is the artist's physical use of the substance in shaping his art, and the greater attunement to the environment from which it is extracted.

Logar's use of site-specific materials transforms bitumen into a representation of the contaminated landscape. In the process the material is fragmented into little particles on a sheet of glass and could also be read as a classified bitumen sample. The bituminous signs of the landscapes and oil infrastructure on glass bring into focus the tension between visibility and invisibility. By assembling these glass sheets into single water tanks, the resource of water is incorporated, and the work creates a pointed image of this specific segment of energy culture.

These six water-tank objects, representing key locations within the Canadian tar-sands industry (tailings ponds, reclamation sites, refineries, pipelines, open-pit mining), form in turn parts of a larger installation.

The water tanks stand as physical reminders of the role played by water in the oil sands—the use of water in extraction practices, the impact of the industry on the state of the Athabasca River, and the pooled consequences of energy desires in the tailings ponds left behind all over northern Alberta.

Fig. 1.3. *Pipeline–Anzac* 2015, Ernst Logar, 2016. Bitumen screenprint on glass, 260 × 360 mm.

Fig. 1.4 (*top*). *Tar Sands: Approaching an Anthropocentric Site*, Ernst Logar, 2016. Installation, PAVED Arts. Water tanks, water samples, three-channel video *Oil on Water* (15:34 min).

Fig. 1.5 (*bottom*). *Tar Sands: Approaching an Anthropocentric Site*, Ernst Logar, 2016. Installation, PAVED Arts. Water tanks, water samples, three-channel video *Oil on Water* (15:34 min).

The tanks are placed on customized, backlit stands around the gallery space in careful relation to a three-channel projection that features video of the Athabasca River, different significant locations of this industry, and a water ceremony of a First Nations community at a tailings pond. The installation made its debut at the PAVED Arts gallery in Saskatoon from 4 November to 10 December 2016, and was curated by artistic director David LaRiviere. Logar's *Oil and Water* shows us that nothing has been left untouched in northern Alberta; this has had significant and ongoing consequences for the Aboriginal communities there.

Untouched bitumen could be imagined as hope because it would signal an entirely different relationship to fossil fuels—a connection to water, to life, and to community that at present is being stripped away as fast as the topsoil that hides bitumen from the ferocious maw of modernity.

—Imre Szeman

CHAPTER 2

Trespassage

Mél Hogan

As Canadians, we cohabitate to varying degrees and proximities with the energy infrastructures on which we depend to move, eat, and communicate. This locatedness has long been informed by the country's climate and topography. And because of where most of us aggregate—as evidenced by power grids, train tracks, highways, oil pipelines, and cell towers—we also tend to define ourselves in relation to our more politically powerful and densely populated neighbors to the south, in the United States. Canadian culture has long been negotiating a cultural space between the heated—and under the Trump presidency, increasingly fascist—affect of the US government and our own northern folklore(s). Who Canadians are now ultimately reflects the climate, a seemingly tepid political temperament.

As settler Canadians, we don't think of our country as having a "South" so much as having different degrees of northernness. The Canadian Far North (*le grand nord*) generally refers to "our" part of the Arctic Circle. The Canadian North officially comprises three territories: Northwest Territories, Yukon Territory, and Nunavut. A very small fraction of the country's population inhabits the north—nonindigenous, Métis, Inuit, and First Nations people are spread out over one-third of Canada's cold, rugged, and vibrant terrain.[1] While thinly dispersed in terms of population, its inhabitants have long thrived there alongside caribou herds, polar bears, and beluga whales.[2] The rest—a very large majority—inhabit a narrow corridor of warmer climate and define themselves by looking northward to a largely unoccupied territory that makes the landscape so majestic. Overall, the nation has been built on a generalized and unspoken "nordicity," but forged more directly by a population that huddles and hovers uncomfortably close to the US border—a thin band of people connecting east and west.[3]

To understand ourselves geographically, then, we might focus on the borders that our bodies claim for us, and those that divide and contain us, as well as those we willfully traverse or sometimes transgress. Looking north, neither water nor landmass alone serves to demarcate the territory of the Arctic.[4] The

North's inhabitants have long considered many areas of the North—coast and land ice—as one space.[5] Inuit have used "place names" to describe different features of the land, water, and ice as a way to convey their historicity, mobility, and occupancy.[6] Now, however, because of global warming, northern ice thaws are requiring a shift in these perspectives. Settler Canadians and Inuit alike may need to reconsider their identities in relation to an increasingly fluid Arctic border—one shared with Norway, Denmark (Greenland), and Russia. Many countries that do not border the Arctic (but which are official "observers" by Arctic Council standards) also have vested interests in reshaping the region, primarily for its economic potential through Arctic drilling and other forms of exploitation and extraction. With climate change, "freeze" and "thaw" in the Arctic are more than metaphors when considering how nations shapeshift and adapt their environments to economic ends, largely informed by energy demands.[7]

To understand the significance of these shifts from ice to water and from territorialization to fluid cohabitation, I propose the concept *trespassage*—a portmanteau word that merges "trespass" with "passage"—as a way to rethink the conceptual entanglements of progress and tradition, where both serve as infrastructures linking time and space.[8] "Trespass" is the idea of going on or through someone's land without permission. It can also point to the violation of a set of personal and social ethics. It's about crossing a line—a border at once physical and emotional, and often bound by law. In the case of infrastructural space, trespassage assumes a fluid border; it's about ice thawing and water forging new and unpredictable currents and flows. "Passage," in contrast, connotes the path that connects one place to another, and also the act of moving through that space. It is transit, movement, motion, and progress.

Joined together as *trespassage*, the term speaks of a dual intention and configuration that can be applied to infrastructure that is deemed national, natural, and public (*très- [very-] passage*), on the one hand, and, on the other, serves decidedly colonial ends (*trespass-age*).[9] "Trespassage" prevents easy dichotomies between fixed and fluid (borders, concepts) and applies to temporal as well as geographical interventions (whose land, at which moments), of which the Arctic is but one example. It also disrupts the current and overwhelming academic focus on concepts of sovereignty and nation-building, and instead transcends a different imaginary from the "Arctic Paradox."[10] In this way, trespassage can be a tool and framework for articulating the space of infrastructure as it sediments itself into the environment, imposes onto its surface, floats above it, or is placed underwater. The term also serves to account for the colonial and imperial ideals and discourses actively reconstituted by way of infrastructural spaces, trajectories, and emplacements.[11] Specifically, here, I use it to get at the complicated ways in which internet infrastructure, and the connectivity it promises, has become de facto betterment—in particular, for energy management, access to health and education, and the preservation of language in rural and northern communities.[12]

"Trespassage" also names the *age of trespass*: the growing surveillance and monitoring capabilities built into our wireline communications infrastructures, as well as the societal control embedded in increasingly pervasive ecological discourses, big-data analytics, cybernetics theory, and machine logics. It is through this concept that I look at the project of laying high-speed internet fiber-optic cable through the Arctic Northwest Passage. The companies responsible for this project—Arctic Fibre, Inc., and later, Quintillion Subsea Holdings, LLC—serve as a case study for the complex, and often false and pervasive, logics and infrastructural arrangements that increasingly form and inform the mediascape of Canada and its identitary delimitations.[13] As stated by environmental media theorists Richard Maxwell and Toby Miller, "communication technologies provide an expanding universe of discourse," and within a globalized and growth-driven capitalist system, the main goal of technology is "to overcome scarcity and bestow the benefits of plenitude though access on every person, all the time."[14] Connectivity, in and of itself, is seen as technological and as inherently good. However, this unified system of telephony, data, mobility, and media also has important energy implications that cool climates, and the idea of the North, quells. Colonial ideals of the North—perhaps best labeled as a "white turn"—are increasingly reinforcing infrastructural logics around both energy consumption and communication. These have been historically tracked together. What I'm calling a "white turn" is essentially the act of looking to the North once again as a "new" space of extraction and exploitation that reinforces and reinstates predominantly white settler logics around land occupation as a "civilizing" mission.

In his 1923 dissertation, Canadian communications scholar Harold A. Innis drew attention to the natural conditions that facilitated and guided the direction of trade and development in Canada. He paid particular attention to the effects of the interconnected flows of lakes and rivers, but also to geological and climatic factors, and flora and fauna habitats that could propel or halt development. At the beginning of his academic career, Innis was interested in studying the completion of the Canadian Pacific Railway (CPR) as well as earlier (failed) quests to get through the Northwest Passage. To him, these were two crucial moments/conquests shaping Canadian identity. While he wrote about the CPR as an economic/historical case study of the industry, he paid little attention to the social conditions that, to this day, are part of the social fabric of the railway system.

Innis focused instead on the infrastructures that facilitated the colonization of North America, what he deemed the "civilization" and "unification" of a nation. The development of the CPR was not framed by politicians at the time as an imperial project, but rather as a necessary imposition serving universal ideals about modernity, improvement, and progress based on trade and commerce as tools for self-governance. Regardless of political posturing, the CPR project was one of trespassage: many Indigenous peoples and their way of life, as well as the life of animals such as the bison, were displaced by the railway, and each year, many animals continue to be killed by passing trains.[15] The tracks provided a system for *passing through* the country, igniting a hotel and tourist industry in the newly delineated national parks (and the creation of protected national parks).[16] While the train tracks were an amazing infrastructural feat, they remain a marker of—and an enduring monument to—colonial and imperial ideals.

An estimated fifteen hundred to seventeen thousand Chinese immigrants worked for the CPR under terrible conditions between 1881 and 1884, their contributions mostly undocumented. These workers' wages were far less than those of their Canadian and European counterparts, even though they were assigned the most dangerous components of the project, dealing with explosives or working on bridges and tunnels. Many Chinese immigrant workers died while working on the construction of the CPR. Then, as now, the infrastructural logics were imbued with the rhetoric of serving a greater (and whiter) good, justifying all repercussions and consequences. Unlike settler colonialism, which relies on persistent structures and narratives—namely, reinforcing the perception that Indigenous-inhabited land is empty or unused—trespassage challenges the colonial power dynamics that frames infrastructure as a positive disruptive force on the environment, inclusive of those who occupy and inhabit those spaces. Like the train, internet cables pass through large territories but connect only certain nodes along the way, from which expansion follows. Having access to the internet is now largely considered a human right,[17] an idea adopted by various Indigenous internet companies as well, equating connectivity with development and with self-governance. Here, "trespassage" serves as a concept that observes (more than it counters or remedies) imaginaries of the North as remote but connected, as cool but hot (fig. 2.1). Concerning the Northwest Passage, Innis wrote about the centuries-old colonial efforts to pathfind, from east to west, from the Atlantic Ocean to the Pacific Ocean through the Arctic Archipelago, in order to create a trade route. The name

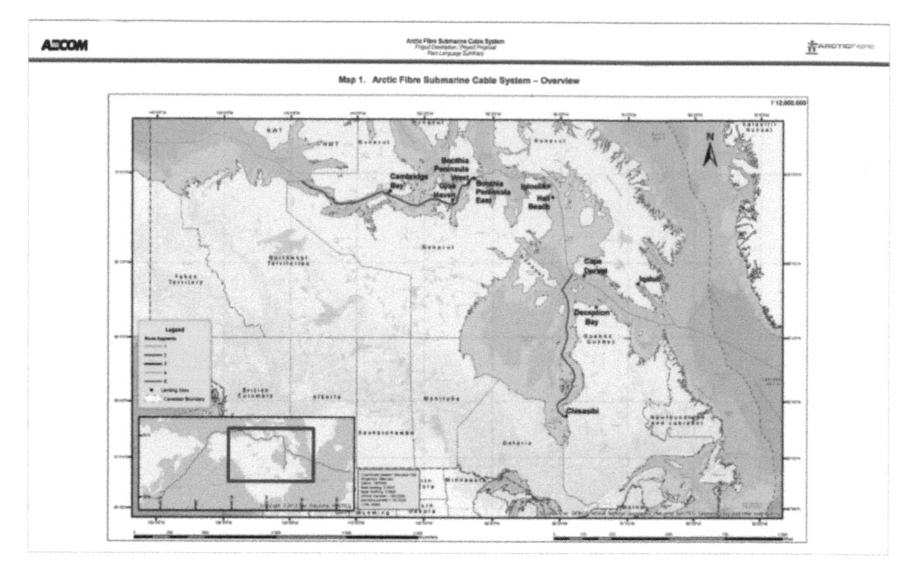

Fig. 2.1. Screen grab of Crystal Cruises through the Northwest Passage, taken from https://www.qiniq.com/company/.

itself, "Northwest Passage," is oriented to Europe. The passage extends 1,450 km, from north of Baffin Island to the Beaufort Sea. Traversing it meant a dangerous voyage, navigating ice fields and dodging tens of thousands of giant icebergs adrift between Greenland and Baffin Island.[18]

Since about 2007, warmer temperatures have enabled a passage relatively free of ice, but only in the late summer.[19] Although it remains a treacherous passage by most measures—with less ice land to detour but with more icebergs (and growlers) to dodge—cruise ships are no longer a rare sight in the Northwest Passage. In 2017, you could book a trip from Anchorage to New York (fig. 2.2). However, when viewed as a trade route, poorly charted sea lanes, liability and insurance concerns, weather conditions, and so on continue to make the passage difficult, though no longer impossible.[20] Today, the Northwest Passage is either an experience for the super-elite or a commercial trespassage from Asia to Europe. There are a few examples of this so-called shortcut that together speak of the shifting vision of the Arctic.[21]

As early as 1969, the SS *Manhattan* set sail through the Northwest Passage to test the waters as a shipping lane for oil transportation from Alaska to Texas. The US ship was assisted by a Canadian icebreaker, but it assumed the passage to be an "international strait" rather than a part of Canada's "internal waters."[22]

QINIQ
The Coolest Hotspot on Earth
ᕿᓂᖅ
Learn more

Fig. 2.2. Screen grab from https://www.crystalcruises.com/voyage/details/northwest-passage-explorer-7320.

In 2013, the large bulk carrier, the *Nordic Orion*, sailed from Vancouver to Finland with a load of 15,000 metric tons of coal. This saved the company four days of travel, compared to its regular route, and an estimated two hundred thousand US dollars in costs.[23] The following year, the cargo ship MV *Nunavik* completed the entirety of the Northwest Passage carrying 23,000 tons of nickel concentrate collected from a Chinese-owned mine near Deception Bay (fig. 2.3). The Northwest Passage is 40 percent shorter than the Suez or Panama Canal routes that would otherwise be taken.[24] This idea of trespassage is rekindled by the elite Crystal Cruises, which brought the world's richest people to experience the rugged fragility of the Arctic.[25] But, more than anything else, these kinds of passages are symbols of innovation, power, and fantasy about the future.[26] Each new attempt to traverse the Arctic speaks to a persistent urge to pass through more quickly than before, to find a shortcut—to save time and money—a knowing trespassage that justifies itself with an efficiency of its own measure.[27]

The vast wilderness and waterways of the Northwest Passage continue to shape global colonial imaginaries: the North as a place to be conquered and developed, but also, for some, a place to be protected and revered. How, then, is the materiality of the internet to be understood through such a rich and important "natured" context? And, more specifically, given these looming infrastructural projects, how might northern communities' political, philosophical,

Fig. 2.3. Map, 2016. Screen grab of website from the Nunavik crossing, Fednav, http://www.fednav.com/en/voyage-nunavik.

and spiritual connection to the land and water be reconciled with the emerging virtual landscape of modernity? The Northwest Passage is increasingly depicted (by transportation and communication industries) as trespassage—a transient place, used only to get elsewhere, but in effect bringing in Western colonial ideas about education and health and the English language. This requires us to question the social and political ramifications of ice thaws, imagine these new routes, and ponder more seriously the ways that land and water are marked, mobilized, and mapped.[28]

For more than a decade, the Northwest Passage has been transforming from impenetrable ice field into navigable waters. Taking advantage of this change in melting landscape, the Toronto-based company Arctic Fibre, Inc. proposed in 2013 a project to replace Telesat's long-standing satellite beams with high-speed internet communications by laying a 15,000 km undersea fiber-optic cable line in the Northwest Passage.[29] In the late 1990s, the government of the Northwest Territories made a deal with a northern-based, Aboriginal-owned company called Ardicom to set up the satellite system. At the time it was seen to be the only system that would sustain and survive the continuous assault of sub-sub-zero conditions.[30]

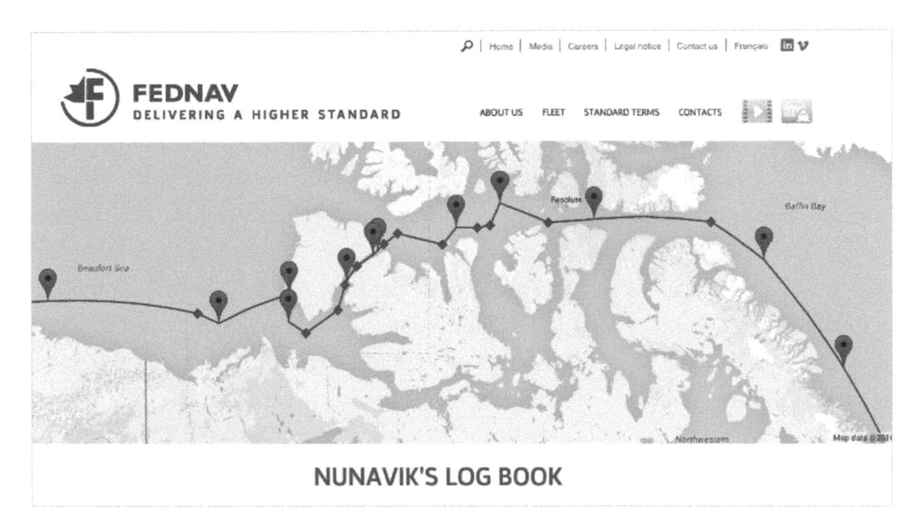

Fig. 2.4. Map taken from the 2013 Arctic Fibre Submarine Cable System Project Description / Project Proposal Plain Language Summary, 3, gcc.ca/pdf/2013-10-23-Non-Tech-Summary.pdf. The map shows the proposed path of the cable system through arctic Canadian waters.

Quintillion Subsea Holdings, LLC is the Anchorage-based company that in 2017 was constructing Phase 1 of the multiphase subsea and terrestrial fiber-optic cable network passing through the Northwest Passage.[31] At this time, it was already no longer evident that a Canadian portion was in the plans. This change of hands—from a Canadian company to an American one—brought on other, subtler shifts as well. Unlike its forerunner, Quintillion is an American company located in the North that it seeks to develop.[32] However, company promotional materials placed no importance on the question of sovereign Indigenous lands, or on the Inuit in particular who dominate the eastern Arctic and who stand to be most affected by the project's future phases, if it were to be realized as originally planned by Arctic Fibre.[33] Instead, they emphasized the potential technical efficiencies of such a network—including quashing the "global demand for redundancy"—and used other jargon that seems to downplay or dismiss how community life is forever changed by infrastructure, as other colonial impositions have done before.[34]

Despite the mention of local communities in the project proposals, this new communications passage is profiting big business above all else—and energy companies especially. It's a privately owned infrastructure, built to last only twenty-five years. The laying of these internet cables has strong metaphorical power: the cables bypass much more than they connect, and mark "progress" into the landscape, while also limiting it to the material and technological, and to a mere quarter of a century. In this way, this project becomes a case for looking at Canada's media landscapes in light of its growing digital communications infrastructure—along with its long legacy of parallelism with the military, energy, transportation, the wilderness, provincial politics, colonialism, and trade routes—as more than a question of technology, but also as political, social, and environmental.[35] The mapping of the cables, in particular, is part of a post–World War II tradition of taking discursive possession of land, water, and natural resources. What gets mapped is never objective but rather a spatial representation and evidence of Western priorities that reinforce one another.[36] Like the train before it, internet cables provide a material line to trace and mark the significance of certain passages over others. Under provisions of the Telecommunications Act of 1993, Arctic Fibre, Inc.'s proposal to Industry Canada was to get licenses to lay cables that would run through the Northwest Passage and connect Asia and Europe. In the company's proposal, it suggested that these cables could become an alternative route, bypassing both the United States and other "problematic terrestrial routes," referring to the surveillance and monitoring issues associated with those in control of material infrastructures. The proposal also makes a case for the "benignness" of its own environmental impacts, while acknowledging that the melting ice of the North is what is making the project possible.[37]

Using the Arctic waterways, the cables would link large cities like Tokyo, Montreal, New York, and London. According to an early proposal, the infrastructure would also provide service to 52 percent of Nunavut's population and include landing sites in Nunavik and northern Quebec.[38] However, connectivity for these communities appeared not to be part of the first phase of development, but rather part of a possible extension of the next two phases. In an expansive territory like Nunavik, largely without roads or electricity (save for stand-alone diesel generators), just how this population of 300,000 inhabitants can gain access to the internet remains a mystery.[39] Nevertheless, the company made its pitch for this northern cable passage based on bridging the Canadian digital divide by "connecting" the North.[40] It boasted of fiber's technological stability, efficiency, and speed for communications, which would see the North finally "cured" of being cut off from the modern world.[41] Assuming that education and health care is made more accessible by high-speed and stable northern internet connectivity, should we be asking what kind of education, and what models for health? Whose ideals are these, and how, if possible, can the internet *not* be a tool or technology that commands colonial compromise?[42]

The company pitch also omitted discussing existing technologies in the

North (such as NorthwesTel and QINIQ), connecting northern communities to one another, and how infrastructures are being repaired and maintained, as well as how important access to and ownership of infrastructure might be, especially for those told that they are being served a life upgrade. It also altogether overlooks the possibility that internet technology may not be the great equalizer it promises to be—but this remains a question for northern communities to determine for themselves, with equal access being central to connecting the North to and among northern communities (rather than to the south). Building from this case, we can extend the critical capacity of "trespassage" to other features of the media system enabled by cables, sensors, and infrastructures of all kinds that are geared to connect the world, but invariably risk disengaging and disassociating the globe from local needs.

Notes

Thank you to the 2016 Banff Energy gang, and to Petra Dolata, Rafico Ruiz, and Shirley Roburn for enriching my understanding of the history of northern telecommunications.

1. See Statistics Canada, *Census Profile, 2016 Census, Statistics Canada Catalogue no. 98–316-X2016001*, Ottawa, February 8, 2017, http://www12.statcan.gc.ca/census-recensement/2016/dp-pd/prof/index.cfm?Lang=E.
2. Sherrill E. Grace, *Canada and the Idea of North* (Montreal: McGill-Queen's University Press, 2002); Peter C. van Wyck, "An Emphatic Geography: Notes on the Ethical Itinerary of Landscape," *Canadian Journal of Communication* 33, no. 2 (June 2008): 171–91; Daniel Chartier, Jean Désy, and Louis-Edmond Hamelin, *La nordicité du Québec* (Quebec: Presses de l'Université du Québec, 2014).
3. L. E. Hamelin, *Nordicité canadienne* (Montreal: Hurtubise HMH, 1975). The North, however, is a contested concept with many fits and starts. In the 1950s the Canadian government displaced seventeen Inuit families (from Inukjuak, in northern Quebec, and Pond Inlet, in Nunavut) to Cornwallis Island and Ellesmere as a way to assert sovereignty over the Northwest Passage. This mission was carried out without the informed consent of the Inuit families, as they were never made aware of the Cold War–inspired motives behind the move, which was intended to position "human flagpoles" to claim that territory for Canada. Herded like animals, these families were labeled with number tags under "Eskimo Identification Canada" and forced to move and to live in the harshest of northern conditions as a means to claim the Arctic. Paul Watson, "Inuit Were Moved 2,000 km in Cold War Manoeuvring," *Toronto Star*, 29 November 2009, https://www.thestar.com/news/insight/2009/11/29/inuit_were_moved_2000_km_in_cold_war_manoeuvring.html. See also Valerie Alia, "Naming in Nunavut: A Case Study in Political Onomastics," *British Journal of Canadian Studies* 19, no. 2 (2006): 247; and Melanie McGrath, *The Long Exile: A True Story of Deception and Survival among the Inuit of the Canadian Arctic* (London: Fourth Estate, 2006).
4. Emilie Cameron, *Far Off Metal River: Inuit Lands, Settler Stories, and the Making of the Contemporary Arctic* (Vancouver: UBC Press, 2015); Wilfrid Greaves, "Thinking

Critically about Security and the Arctic in the Anthropocene," *Arctic Institute*, 22 March 2016, http://www.thearcticinstitute.org/thinking-critically-about-security-and-the-arctic-in-the-anthropocene; Rafico Ruiz, "Arctic," in *Fueling Culture: 101 Words for Energy and Environment*, ed. Imre Szeman, Jennifer Wenzel, and Patricia Yaeger (New York: Fordham University Press, 2017).

5. Mia M. Bennett, Wilfrid Greaves, Rudolf Riedlsperger, and Alberic Botella, "Articulating the Arctic: Contrasting State and Inuit Maps of the Canadian North," *Polar Record* 52, no. 6 (2016): 630–44.
6. Claudio Aporta, "The Trail as Home: Inuit and Their Pan-Arctic Network of Routes," *Human Ecology* 37 (2009): 131. See PanInuit Trails Atlas: http://www.paninuittrails.org/index.html?module=module.about.
7. Petra Dolata, "How 'Green' Is Canada's Arctic Policy? The Role of the Environment and Environmental Security in the Arctic," *Zeitschrift für Kanada-Studien* 32, no. 2 (2012): 65–83, http://www.kanada-studien.org/wp-content/uploads/2012/12/04_Dolata.pdf.
8. Shiri Pasternak, "How Capitalism Will Save Colonialism: The Privatization of Reserve Lands in Canada," *Antipode* 47, no. 1 (2015): 179–96.
9. "Very, 1815, from French très, from Old French tres, from Latin trans 'beyond' (see trans-), later 'very' (cf. Old Italian trafreddo 'very cold')"; http://www.dictionary.com/browse/tres.
10. See Cristine Russell, "The Arctic Paradox Poses Questions about Sustainable Development," *Scientific American Blog*, 5 November 2015, https://blogs.scientificamerican.com/guest-blog/the-arctic-paradox-poses-questions-about-sustainable-development/.
11. Ashley Carse, "The Anthropology of the Built Environment: What Can Environmental Anthropology Learn from Infrastructure Studies (and Vice Versa)?" *Engagement*, 17 May 2016, https://aesengagement.wordpress.com/2016/05/17/the-anthropology-of-the-built-environment-what-can-environmental-anthropology-learn-from-infrastructure-studies-and-vice-versa; Zoe Todd, "Commentary: The Environmental Anthropology of Settler Colonialism, Part I," *Engagement*, 11 April 2017, https://aesengagement.wordpress.com/2017/04/11/commentary-the-environmental-anthropology-of-settler-colonialism-part-i/.
12. "Bridging the Gap: Community Engagement, Content, and Connectivity in the North," workshop organized by University of Alberta North, 20 April 2017.
13. Sarah Rogers, "News: Fibre Optic Cable Headed for Canadian Arctic?" *Nunatsiaq Online*, 24 January 2012, http://www.nunatsiaqonline.ca/stories/article_print/25779; Rob Powell, "Quintillion Buys Arctic Fibre," *Telecom Ramblings*, 18 May 2016, http://www.telecomramblings.com/2016/05/quintillion-buys-arctic-fibre.
14. Richard Maxwell and Toby Miller, "The Environment and Global Media and Communication Policy," in *The Handbook of Global Media and Communication Policy*, edited by Robin Mansell and Marc Raboy (Malden, MA: Wiley-Blackwell, 2011), 467.
15. "While trains are now the leading cause in human-induced grizzly bear mortality in Banff and Yoho national parks, there is renewed optimism for reducing the risks to grizzly bears and other animals in the park that regularly travel along the railway." See http://www.pc.gc.ca/eng/pn-np/mtn/conservation.aspx.
16. Before cars became the norm for middle-class family travel in the 1960s, train

travel was most popular among vacationers. See http://www.cpr.ca/en/about-cp/our-history.

17. See Jeff Lagerquist, "UN Council Says Blocking Internet Access Violates Human Rights," *CTV News*, 4 July 2016, http://www.ctvnews.ca/sci-tech/un-council-says-blocking-internet-access-violates-human-rights-1.2972911. See also David Rothkopf, "Is Unrestricted Internet Access a Modern Human Right?" *Foreign Policy*, 2 February 2015, https://foreignpolicy.com/2015/02/02/unrestricted-internet-access-human-rights-technology-constitution/.
18. "Northwest Passage: Trade Route, North America," *Encyclopedia Britannica*, http://www.britannica.com/place/Northwest-Passage-trade-route.
19. Michael Byers, "Arctic Tourism Heating Up as Northwest Passage Melts," *CBC News*, 24 August 2012, http://www.cbc.ca/news/canada/north/arctic-tourism-heating-up-as-northwest-passage-melts-1.1141578.
20. Whitney Lackenbauer and Adam Lajeunesse, *On Uncertain Ice: The Future of Arctic Shipping and the Northwest Passage* (Calgary: Canadian Defence & Foreign Affairs Institute, December 2014), http://adamlajeunesse.com/uploads/3/4/9/1/34912685/on_uncertain_ice.pdf; Hugh Stephens, "The Opening of the Northern Sea Routes: The Implications for Global Shipping and for Canada's Relations with Asia," SPP Research Paper 9, no. 19 (May 2016), https://ssrn.com/abstract=2808661.
21. Michael Byers, *Who Owns the Arctic? Understanding Sovereignty Disputes in the North* (Madeira Park, CA: Douglas & McIntyre, 2014); Jordan Pearson, "Arctic Cruises for the Wealthy Could Fuel a Climate Change 'Feedback Loop,'" *Motherboard*, 15 August 2016, http://motherboard.vice.com/read/arctic-cruises-for-the-wealthy-climate-change-feedback-loop-crystal-serenity-northwest-passage.
22. Byers, "Arctic Tourism Heating up as Northwest Passage Melts." See Ross Coen, *Breaking Ice for Arctic Oil: The Epic Voyage of the SS Manhattan through the Northwest Passage* (Fairbanks: University of Alaska Press, 2012); Adam Lajeunesse, *Lock, Stock, and Icebergs: A History of Canada's Arctic Maritime Sovereignty* (Vancouver: UBC Press, 2016); Byers, *Who Owns the Arctic?*; and Michael Byers, "Why Trudeau Should Move Now to Safeguard the Northwest Passage," *Globe and Mail*, 16 May 2018, http://www.theglobeandmail.com/opinion/why-trudeau-should-move-now-to-safeguard-the-northwest-passage/article31382232/.
23. John McGarrity and Henning Gloystein, Reuters, "Northwest Passage Crossed by First Cargo Ship, the Nordic Orion, Heralding New Era of Arctic Commercial Activity," *National Post*, 27 September 2013, http://news.nationalpost.com/news/canada/northwest-passage-crossed-by-first-cargo-ship-the-nordic-orion-heralding-new-era-of-arctic-commercial-activity. Chester Dawson, "Cargo Ship Carves a Path in Arctic Sea," *Wall Street Journal*, 25 September 2013, http://www.wsj.com/articles/SB10001424052702304526204579097582965775144.
24. However, according to Loadstar UK, "the Panama and Suez canals could be the next institutions to be affected by the ongoing crisis in the container shipping industry, according to new analysis from Africa ports analyst and monitoring service portoverview.com." See "Canals Feel Ripples of Container Shipping Crisis as Vessels Go the Long Way Round," *G Captain*, 23 February 2016, http://gcaptain.com/canals-feel-ripples-of-container-shipping-crisis-as-vessels-go-the-long-way-round/.
25. Pearson, "Arctic Cruises for the Wealthy Could Fuel a Climate Change 'Feedback Loop.'"

26. See "Nunavik's Log Book/Fednav," http://www.fednav.com/en/voyage-nunavik; and "First Arctic Cargo Shipped through the Northwest Passage/Fednav," http://www.fednav.com/en/media/first-arctic-cargo-shipped-through-northwest-passage-0.
27. Margaret Kohn, "Colonialism," in *The Stanford Encyclopedia of Philosophy*, edited by Edward N. Zalta, Spring 2014 edition, http://plato.stanford.edu/archives/spr2014/entries/colonialism/; Judy Wajcman and Nigel Dodd, *The Sociology of Speed: Digital, Organizational, and Social Temporalities* (Oxford: Oxford University Press, 2017).
28. Bennett, Greaves, Riedlsperger, and Botella, "Articulating the Arctic: Contrasting State and Inuit Maps of the Canadian North"; Rafico Ruiz, "Locative Media and the Production of Georesources: A Pan-Arctic Spatial Data Infrastructure," *Spheres: Journal for Digital Cultures*, June 2016, http://spheres-journal.org/locative-media-and-the-production-of-georesources-a-pan-arctic-spatial-data-infrastructure/.
29. "Telesat Proposes $40 Million Investment in Arctic Infrastructure to Expand Broadband Services in the North," *Cision News Wire*, http://www.newswire.ca/news-releases/telesat-proposes-40-million-investment-in-arctic-infrastructure-to-expand-broadband-services-in-the-north-509539801.html. See also "ACIA Report: Insight into the Economics of 'Broadband,'" 2011, http://www.aciareport.ca/chapter8.html.
30. As of 18 May 2016, however—and in the course of writing this piece—the Arctic Fibre, Inc. website went down, its URL (http://arcticfibre.com) redirecting to the Quintillion Subsea Holdings, LLC site (http://quintillionnetworks.com), which then also went down in late August 2016 (to become http://qexpressnet.com). It's unclear what these shifts mean, but they suggest a kind of virtual instability that foreshadows the consequences of hasty development. Is the melting land ice not making us all slow down rather than forge ahead? See https://www.qiniq.com/wp-content/uploads/2015/07/Wired-1997–11–05.pdf.
31. As of 18 May 2016, Arctic Fibre, Inc. operates as a subsidiary of Quintillion Subsea Holdings LLC. See http://www.bloomberg.com/research/stocks/private/snapshot.asp?privcapId=183396880.
32. The company is backed by Cooper Investment Partners, a private investment firm based in New York.
33. This changed throughout the writing of this paper, with more visuals becoming part of the PR materials, drawing attention to the northern inhabitants.
34. See http://quintillionnetworks.com/?page_id=25 and http://qexpressnet.com.
35. Susan Leigh Star, "The Ethnography of Infrastructure," *American Behavioral Scientist* 43, no. 3 (November 1999): 380; Rafico Ruiz, "Arctic Infrastructures: Tele Field Notes," *communication +1* 3, no. 1 (2014): 1–25; Nicole Starosielski, *The Undersea Network: Sign, Storage, Transmission* (Durham, NC: Duke University Press, 2015).
36. See http://www.paninuittrails.org/index.html?module=module.about as a counter-example.
37. See Arctic Fibre, Inc., *Arctic Fibre Submarine Cable System Project Description /Project Proposal Plain Language Summary*, 2013, http://www.gcc.ca/pdf/2013–10–23-Non-Tech-Summary.pdf.
38. For its connection to Iqaluit, the company proposes landing the cable at a spot near the access road that leads to the old Hudson's Bay Co. buildings in Apex. See http://www.nunatsiaqonline.ca/stories/article/65674photo_a_little_bit_of_fibre_goes_a_long_way.
39. Stéphane Champagne, "Brancher les villages du Nunavik," *La Presse*, 25 April 2014,

http://affaires.lapresse.ca/portfolio/developpement-du-grand-nord/201404/25/01-4760804-brancher-les-villages-du-nunavik.php.

40. Rob McMahon, "Creating an Enabling Environment for Digital Self-Determination," in *Media Development: Indigenous Media and Digital Self-determination*, edited by Philip Lee (Toronto: World Association for Christian Communication, 2014); Rob McMahon, Susan O'Donnell, Richard Smith, Brian Walmark, Brian Beaton, and Jason Simmonds, "Digital Divides and the 'First Mile': Framing First Nations Broadband Development in Canada," *International Indigenous Policy Journal* 2, no. 2 (2011): 1–15.
41. See Northern Communications & Information Systems Working Group, *A Matter of Survival: Arctic Communications Infrastructure in the 21st Century*, Arctic Communications Infrastructure Assessment Report, www.aciareport.ca.
42. According to Sarah Kester: "The Inuktitut word for 'internet,' 'ikiaqqijjut,' is often translated as 'the tool to travel through layers.'" "When the World Went Online, Inuktitut Followed," *Meet the North*, 2016, https://www.meetthenorth.org/2016/09/when-the-world-went-online-inuktitut-followed/.

CHAPTER 3

The Ocean and the Cloud: Material Metaphors of Hidden Infrastructure

Jayne Wilkinson

What lies between the ocean and the cloud? Between the sea and the sky? What is so fascinating about the vast and vertically organized spaces between the lightness of the stratosphere and the intense pressure of the deepest bodies of water? We frequently gaze upon the elements of water and air, poetically naming their capacities to affect us, but to what end? Oceans are weather regulators, their currents circulating in response to the pull of gravity and the constant axial turn of the Earth. The clouds reply, continually forming, dissolving, and re-forming as patterns written into the sky. More than any terrestrial ecosystem, ocean currents control the climate and determine the weather events that have become ever more unpredictable. In an era of energy crises and climate collapse, our attention has turned to the oceans with rapture.

In film and video, in the spew of images across our daily data and news feeds, in narrative and literary fiction, and in the familiar shipwrecks and ocean storms depicted in the oil paintings of another era, the ocean is as familiar to those landlocked as it is to those whose livelihoods depend upon it. As a concept, the ocean has representational boundaries that are both singular and metaphorical—"the ocean"—while the sites of specific oceans, seas, and shorelines are varied, diverse, and multiple. Likewise, visual representations of the atmosphere are frequently bound by the beauty and sublime tumult of storms, sun, and stars, written in images that belie the material existence of satellite infrastructure, commercial air and cargo transit, pollutants, and other forms of industrial and military aerial occupation. The "cloud" became the tech industry's marketing device par excellence precisely because it suggested that digital communication was untethered from the material realities of the Earth, playing on our tendency to regard the atmosphere as an ethereal, nowhere space.[1] But the "cloud" does tangibly exist in multiple, physically specific sites, including the miles of fiber-optic cables that encircle the globe along the oceans' floors. The ocean-as-network may yet ascend in global consciousness, as sea levels rise and plastics proliferate; like the overdetermined metaphor of cloud computing, in the ocean, metaphor and materiality begin to blur. Throughout this chapter, both ocean and sky function as idiosyncratic signifiers: fragments of singular metaphors based upon elemental spaces but writ large through multiply differentiated sites of planetary communications infrastructure. Both conjure images and narratives that rely on metaphorical thinking, and upon a horizon line that is endless, unstable, and continuously connecting one to the other, sea to sky. Yet it is precisely such evocative habits of mind that prevent us from seeing the fiber-optic cables strung across the ocean floor or the satellites held in geostationary orbit for their true functions: as signifiers of global media infrastructure. Rather than simply reveal these signifiers as such, I aim to bridge divergent modes of aesthetic address so as to demonstrate how metaphorical thinking disguises the realities of networked space. Communication networks have long been embedded within the Earth's atmosphere and its oceanic depths, and while the sky and the sea remain evocative muses, if the climate crisis is to be addressed fully, there is urgency in seeing the planetary-scale networks we've hard-wired into our environments.

Open Metaphor

In her poetic, mystical descriptions of the blues of sea and sky, writer Rebecca Solnit situates readers within the shifting color fields of atmospheres both aerial and oceanic:

> Water is colorless—shallow water appears to be the color of whatever lies beneath it, but deep water is full of this scattered light, the purer the water the deeper the blue. The sky is blue for the same reason, but the blue at the horizon, the blue of land that seems to be dissolving in the sky, is a deeper, dreamier, melancholy blue, the blue at the farthest reaches of the places where you see for miles: the blue of distance.[2]

This blue of distance, this symbolically and spatially assured color, is a deeply held metaphor for the longing one feels looking across the surface of the sea. From the perspective of the terrestrial human, we are unable to see its contours or measure its depths, but we remain certain that the horizon enfolds within it the simultaneous capacity for loss and desire. We see only the surface, mirroring the blue of the sky above and obscuring the mysteries of what is below. The horizon becomes an operative mediator, a nonspace, a line that is always moving, just ahead of us, just slightly out of view.

Despite the apparatuses we build—boats, scuba gear, bathyspheres, submarines, rockets, satellites, spacecraft—both ocean and atmosphere remain distant, making the sea an easy dumping ground for the mountains of plastic produced by half a century of consumerism and the sky a receptacle for chemical pollutants. And, like the waters of the world whose currents circulate an increasing amount of plastic particulates, the sky is full of a mix of dead satellites and space debris orbiting the Earth in perpetuity, their gravity-tethered bodies the remnants of just a few decades of space exploration.[3] The environmental reasons to care about the biological viability of sea and sky seem obvious enough, but perhaps less obvious is the recognition that oceans act as a medium for nearly all our contemporary communication systems. *The oceans are networked spaces.* While garbage gyres, rising sea levels, and species extinction are the more familiar—or simply more visible—examples of human impact, the oceans also contain a massively distributed, precisely mapped, international communication infrastructure: the submarine cables that link the internet and drive nearly all of its traffic. This network, an expansion of the nineteenth-century telegraph cable network, now contributes to one of the most singular energy draws on the planet.[4] The powerful metaphors that articulate our desires around longing, distance, and infinite space and drive our aesthetic understanding of the ocean no longer seem relevant. Instead of being viewed through metaphors of distance, danger, or the unknown, the ocean should be more clearly recognized as the embodiment of a tangible, powerful, and literal network, one capable of collapsing vast distances through near-instantaneous video-, image-, and text-based messaging, rendering the seas less a symbol of division and more a space of interconnection.

Against the persistent symbolism of distance and infinity, and as a powerful medium of our networked global imaginary, the oceans contain sites of connectivity that run deep below their surfaces and make landfall in various, often invisibly protected landing sites. Literalizing a ground of sociability and communication operative under a figurative ocean of invisibility, this dialectic is emblematic of the many entanglements holding our knowledge of the world to the same material systems that threaten it. Technology theorist Paul N. Edwards, in a study of the birth of networks, articulates metaphor as a major form of representation, one that is as important as the material form of the machine. His account of the closed world of Cold War–era cybernetics and geopolitics argues that metaphor is a useful tool around which to organize theories and structure material relations. Far more than a rhetorical or linguistic device, "metaphor is part of the flesh of thought and culture, not merely a thin communicative skin. Therefore the politics of culture is, very largely, a politics of

Fig. 3.1. Camera-phone image taken by author, 2016.

metaphor, and an investigation of metaphor must play an integral role in the full understanding of any cultural object. The mind is such an object, and the computer is such a metaphor."[5]

What happens if we replace mind with ocean and computer with space in this last analogy, if the ocean is the object and space the metaphor? As a cultural object, the ocean itself is overused *as* a metaphor, but if we consider oceans in their material contexts—polluted, acidifying, depleted ecologies containing a vast planetary network that supports continued environmental destruction—the investigation into metaphor that Edwards argues for might account for a common understanding of oceans as space, as networked space. Any type of infrastructure is notoriously difficult to see; in order to be useful, its visuality must be fragmented while its power is totalizing. Across contemporary communication networks, each human operator is only one node in a massively distributed system, one that increasingly relies upon nonhuman labor and artificial intelligence. Only through the power of a catastrophic weather event—or the force of a general strike or labor cessation, one equally catastrophic to capital—is the global connectivity of the supply chain revealed. The slow delivery of consumer products across the surface of the sea is perhaps a more appropriate image to evoke the (coming) complexities of the technological sublime embodied in and by cloud computing, more so than the way "the cloud" is often portrayed.

Only recently have documentaries, maps, and images become more available to the public, revealing and deciphering the global infrastructure that supports the internet through its many data centers, cable systems, network hubs, and corporate campuses. The submarine cable network has its historical grounding in the distribution of telegraph and telephone networks beginning in the nineteenth century, yet with the advancement of satellite technologies and the corporate marketing of the singular "cloud" as the site where vast quantities of private data are stored it seems remarkably strange that our digital communication networks should operate with lightning speeds from the depths of the ocean—and not through the skies. The very existence of such a network stretches one's imaginative capacities. A single fiber-optic cable is typically the diameter of a human hair; it is then bundled together with hundreds of others, encased in several protective layers, spliced into lengths that are thousands upon thousands of miles long, then spooled into the sea by purpose-built ships that will monitor their slow sinking and eventually permanent placement on the ocean floor.[6] Recall that it is only the first 200 m of ocean depth that holds light. There are very few solid "floors" to the ocean, and most of the ocean bottom is not solid at all but is a muddy, sludgy accumulation

Fig. 3.2. Camera-phone image taken by author, 2016.

of falling organic debris, remnants of life that never quite harden. The ability to withstand the immense pressure of these spaces is what makes life at oceanic depths so unlike human life, so alien. This unusual and unhuman space is where undersea cables are laid, allowing us to send messages through light signals bounced along fibers and across the globe through its darkest, densest, and surely strangest spaces.

If this reads as futuristic—all this light becoming information in the ultimate darkness of the oceanic benthic zone—it's because it is implausibly difficult to imagine. Yet once mapped, once made visible as a totality, the cable systems appear all too real, recalling the trade routes that facilitated the colonial expansion of a previous century, now sunk from the surface but still

Fig. 3.3. *Submarine Cable Map* (2015) by TeleGeography. A map showing submarine fiber-optic cables connecting global internet networks, with elements of medieval and Renaissance cartography. Courtesy TeleGeography / telegeography.com.

charting the fastest, most direct route between centers of power. The American telecom mapping company, TeleGeography, produced a map in 2015 using a somewhat tongue-in-cheek style of Renaissance cartography, complete with monsters at sea, to describe the global communication systems of fiber-optic connectivity. It reminds one of the colonial impetus of such a global project, made clear by the references to maps of a previous era, used primarily for the purposes of charting oceanic routes to new colonies, enabling territorial expansion and resource extraction for the European powers.

Closed World

The networking of oceanic space is a project not unique to the digital era. It was achieved in 1858 with the first telegraph transmission along a transatlantic cable laid between the west of Ireland and the easternmost point of Newfoundland. This linked the United States with the United Kingdom for

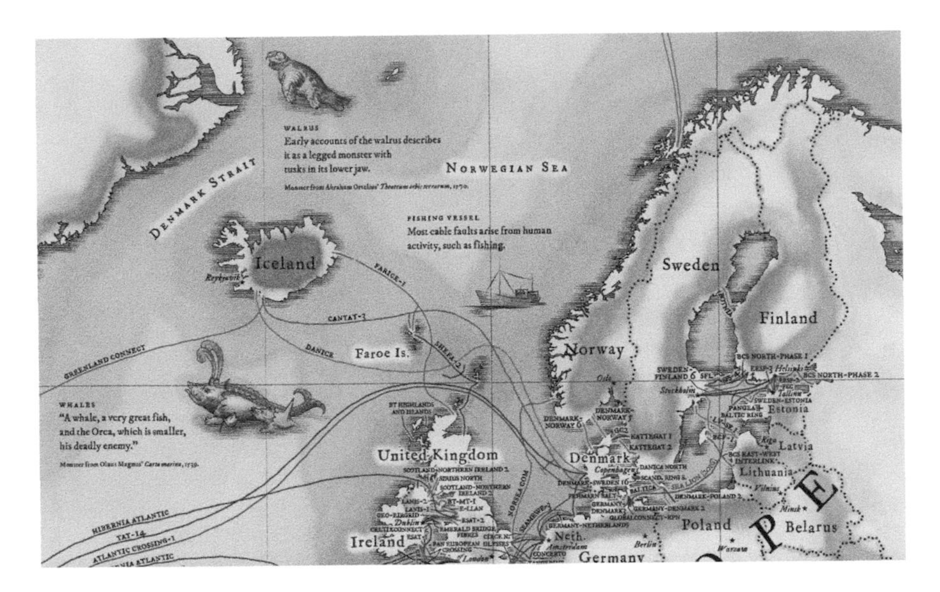

Fig. 3.4. *Submarine Cable Map* (2015) by TeleGeography. Map depicts 299 active and planned cables, as of 2015. Courtesy TeleGeography / telegeography.com.

the first time, allowing for the exchange of messages within minutes through the subsurface cables, instead of the weeks required by surface-crossing ships delivering the post.

In *The Undersea Network* (2015) media theorist and cultural geographer Nicole Starosielski considers not only the historical trajectory from copper to fiber-optic cables, but also the paradoxes of visibility around this network of critical infrastructure. While infrastructure is typically produced to be hidden and inaccessible to the public—an impulse embraced by corporations that shield the public from the often dangerous and low-paid maintenance labor upon which logistics and global supply chains rely—submarine cables require the production of an appropriate kind of visibility in order to be properly maintained. Instead of displacing the cable network as invisible infrastructure, Starosielski argues that we should read submarine cables as a material system that balances requirements for both a signed and mapped visibility—required so as to avoid cables being severed accidentally by anchors, fishing nets, or pleasure craft—and the simultaneous production and maintenance of an invisibility to those who would oppose their presence.[7] At sea, cable locations must be carefully charted so as to allow navigation; they might be subsurface

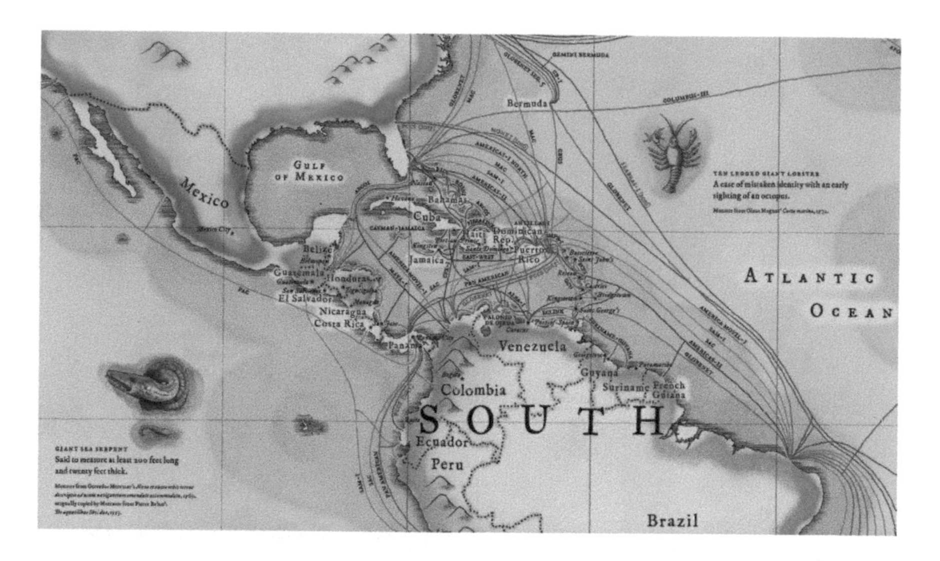

Fig. 3.5. *Submarine Cable Map* (2015) by TeleGeography. Map depicts 299 active and planned cables, as of 2015. Courtesy TeleGeography / telegeography.com.

but they are far from invisible. Yet, once divers pull them onto the land and secure their connections, cables must be thoroughly embedded within the landscape so as to be invisible to potential attack. In California they come ashore hidden as "ecologically sensitive preserves," so as to avoid detection by the environmental activists populating the coast. In Cornwall, the largest landing site in the United Kingdom, the more bucolic option is to disguise them as a family farmhouse.[8] Starosielski notes that by "tracking the cultural production of these visibilities, one can see the material routes of infrastructures supporting 'dematerialized' digital networks, and come to understand the ways in which global systems surface and take shape in relation to local spatial politics."[9] Consonant with techniques for familiarizing the technologies of colonial trade upon which today's communication system is built, the undersea network is felt directly through the local organization of space and the matrices of infrastructure.

The nineteenth century saw massive growth in international infrastructure, but it was corporations, and organization between countries, that produced infrastructure *outside* of the sovereign borders of the nation-state. Architectural theorist Keller Easterling explores the relationship between state sovereignty and infrastructure through what she terms "extrastatecraft," a portmanteau term describing the often undisclosed activities outside

Fig. 3.6. California cable station, location unknown. From Nicole Starosielski, *The Undersea Network* (2015). Courtesy of the author.

of, in addition to, and sometimes even in partnership with statecraft itself.[10] Information resides not only within the technological apparatus—here, the cable—but also in the space that is constructed for and around it. The internet's infrastructural, oceanic matrix operates in such a way: literally offshore and distributed, predominantly situated in international waters, a production of negotiated visibility and a collision of juridical powers. Easterling deploys water metaphorically not only to suggest a sensorium of information carried within the ocean depths, "the space in which we're swimming," but also to recall the primordial sea as a place of generation. We take little notice of the infrastructure upon which we regularly rely because it engulfs us and contains us, like the primordial ocean from which life emerged.

(A textual digression: in the 1967 story *Blood, Sea*, Italo Calvino describes how the hollow cavities of the earliest life-forms made biological life coexistent with the salt water of the sea. As life evolved, the seas moved inward, and blood—the biological system for circulation—became the sea inside us: "the sea where living creatures were at one time immersed is now enclosed within their bodies."[11] This is a rich metaphor for understanding the ways in which we have monopolized the seas to reinsert *our* circulatory networks—not our

Fig. 3.7. Porthcurno Cable Station, Cornwall, England. Courtesy of Bill Burns / Atlantic-Cable.com.

biological networks but the material supports of a system of communication, accumulation, and capital exchange.)

If the seas connect us materially at a cellular level, metaphors drawn from the sky propose communication through an entirely dematerialized space. With the omnipresence of mapping and searching apps, Google Earth's monopoly on satellite imagery, and the pleasures derived from browser extensions like Land Lines,[12] the visuality of the digital Anthropocene is a view from space to Earth through the cloud. In this historical trajectory, photography, weather monitoring, and mapping all intersected for the first time with the advent of aerial photography in the early part of the twentieth century. Following its frequent deployment for military reconnaissance during World War I, aerial photography became an important tool for the National Oceanic Survey (US). Relying upon a system of highly developed black-and-white aerial photographic techniques, the survey used imagery to supplement its fieldwork in order to control and chart shorelines, rocks, reefs, and other natural features.

After World War II, technological advancements in color film became useful for interpreting images of the water from the air. Remarkably, from the early 1960s, the "color photogrammetric system" used color glass plates for direct viewing in stereoscopic plotting instruments. Map and chart compilers

Fig. 3.8. NASA satellite image of hurricane formation over the Pacific, 20 January 1990. NASA ID: S32-80-036. Courtesy of NASA.gov.

would simply "see" and interpret aerial photographs, translating the visual data for use in coastal charting, coastal engineering, mineral exploration, pollution studies, land-use mapping, agriculture, geology, and forestry.[13] When the first satellite was launched by the Soviet space program in 1957, it sparked an intense space race as the Cold War required the Americans to prove the strength of their engineers. Not content to cede power to Soviet scientists, according to NASA, Americans marked 12 July 1962 as "*the day information went global*," with the launch of Telstar, the world's first active communications satellite.[14] Two days later, the United States relayed its first transatlantic television signal. Like the first telegraph signal a hundred years before, this

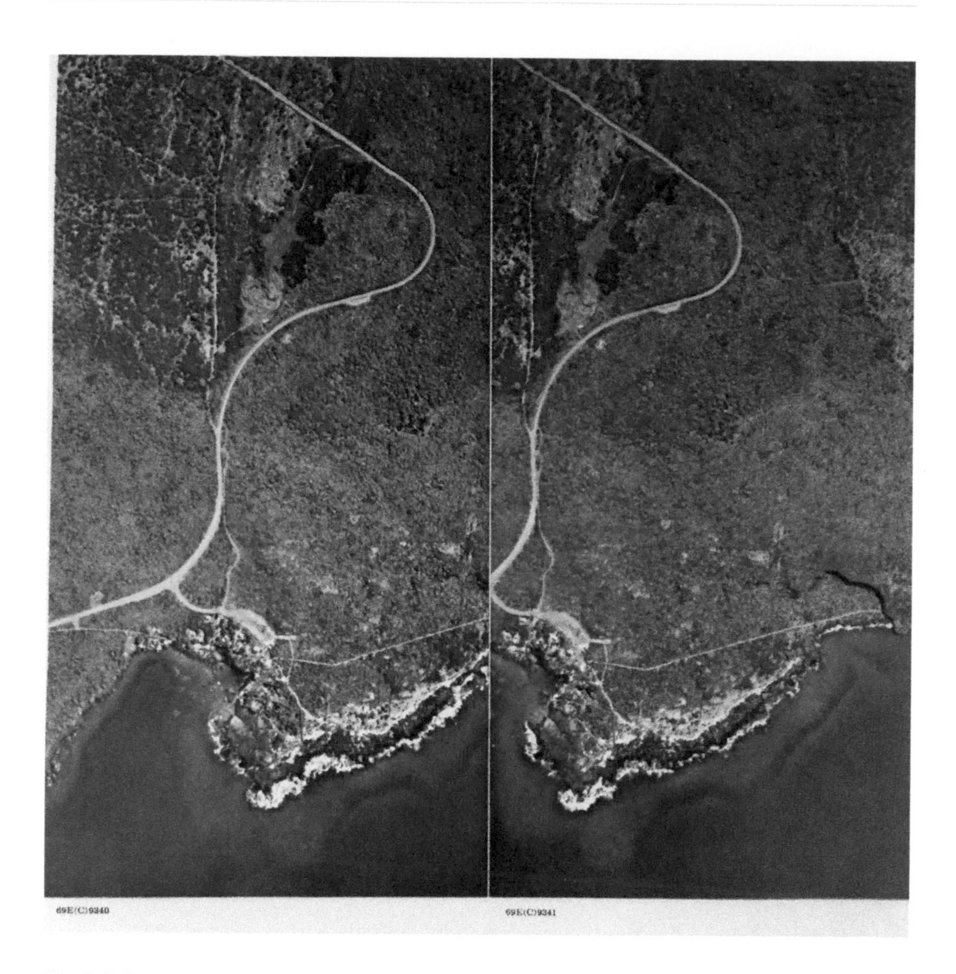

Fig. 3.9. Scan showing a portion of the west coast of Hawaii, an example of photogrammetry. From Harland R. Cravat and Raymond Glaser, Color Aerial Stereograms of Selected Coastal Areas of the United States (Rockville, MD: US Department of Commerce, National Oceanic and Atmospheric Administration, and National Ocean Survey, 1971).

technological feat was used symbolically to cement America's role as a global superpower, and the first images sent were of President John F. Kennedy, the American flag, and Mount Rushmore. Although a planetary-scale communications network had been in place for over a century, it was the American Cold War obsession with space travel and cybernetics that gave birth to the "cloud" as the metaphor for a dematerialized, wireless, smooth, light communications ideology that remains firmly entrenched.

Today, we are wound ever more tightly within a system of control and

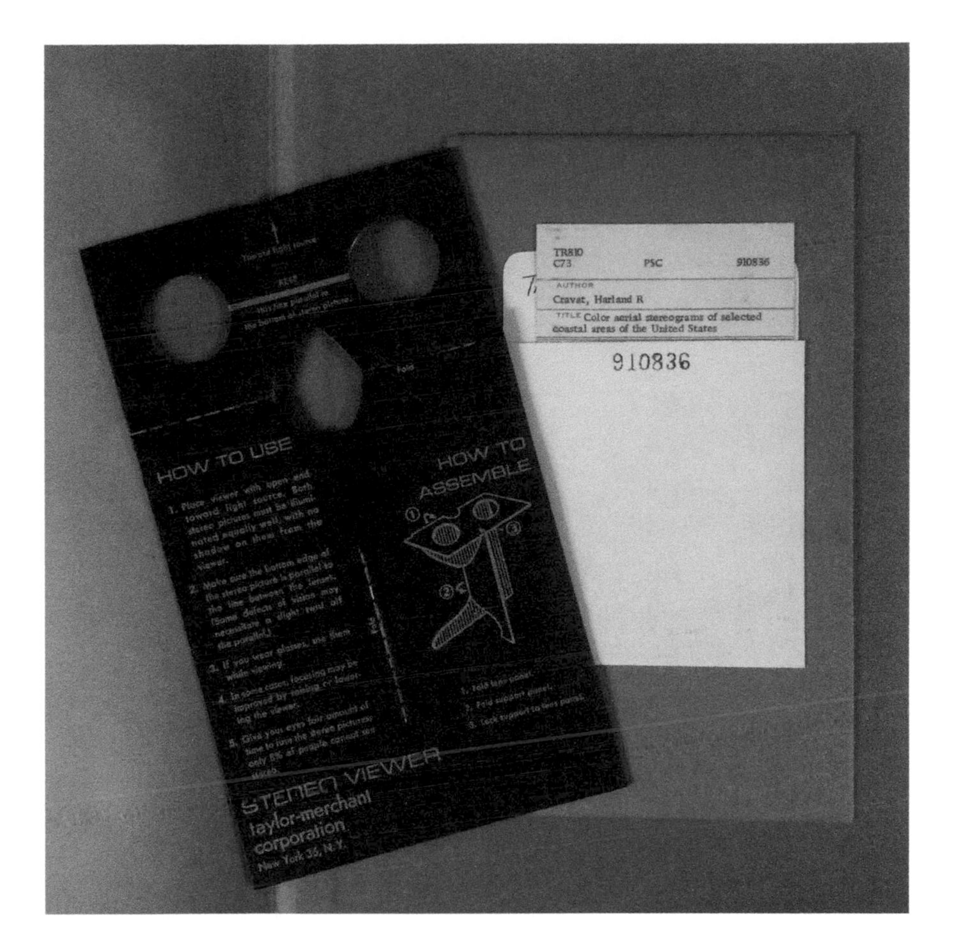

Fig. 3.10. Designed to demonstrate these techniques to a general audience, *Color Aerial Stereograms of Selected Coastal Areas of the United States* (1971) included a fold-out viewing device so readers could view the stereoscopic images and themselves search for variations in land and sea.

computation, while metaphors of space, distance, and vastness continue running in the background. In Edwards's conception, this is a "closed world,"

> a radically bounded scene of conflict, an inescapably self-referential space where every thought, word, and action is ultimately directed back toward a central struggle. It is a world radically divided against itself. Turned inexorably inward, without frontiers or escape, a closed world threatens to annihilate itself, to implode.[15]

Ours is an Earth divided against itself, with no uncharted terrain, with no possibility of escape, and where every action is returned toward the central struggle of survival on a depleted and underresourced planet. The oceans and the sky are now metaphors for a new kind of closed world—not the geopolitical stalemate of the Cold War, but the environmental and ecologically disastrous endgame of planetary collapse.

What can the ocean say about the sky, and vice versa? What can a surface tell us about a depth? What information is hidden within an image? And how can a metaphor be materially grounded? The examples that follow respond to these lines of inquiry, suggesting the vagaries of distance between what is knowable and unknowable in the opaque, incomplete aesthetics of water and sky.

American artist Roni Horn's long engagement with water and weather, as intertwined subjects of her practice, asks her viewers to consider their own personal relationship with water. She often uses literature and writing as a form of visual art, and her monologue *Saying Water* (2012) is one of her most direct addresses to water. She asks these questions:

> When you say water what do you mean? When you say water are you talking about the weather? Or yourself? When you see your reflection in water, do you recognize the water in you? The deserts of our future will be deserts . . . of water.[16]

She returns again and again to the notion that we are water; our subject positions cannot be untangled from it, and the desire to see ourselves mirrored in it is not hidden in those self-same reflective capacities. The weather is water, the Earth is water, our bodies are water; its omnipresence signifies both living and dying in a piercing simultaneity. In Horn's renowned series of photolithographs, *Still Water (The River Thames, for Example)*, she depicts the ever-changing surface of the River Thames in dark, mottled tones. Each image focuses on a small area of the river so as to highlight the dramatically different qualities of light, and the reflective properties of water as it becomes both transparent and opaque. Interspersed across the photos are tiny footnotes, the numbers embedded within the image and nearly imperceptible corresponding notes included below. The notes draw the viewer into a literal reading of the work, collapsing the surface of the water with the depth of a narrative that is subjective, associative, and frequently traumatic. One such note reads: "The opacity of the world dissipates in water. Black water cannot dissipate the opacity of the world."

Fig. 3.11. Image of the Telstar 1 Satellite. Courtesy Bell Labs / NASA.gov.

Roni Horn's surface abstractions belie narratives of loss and tragedy but do little to specify for whom such losses persist. For Christina Sharpe, a theorist who has produced perhaps one of the strongest metaphors of oceanic thinking in recent scholarship, the opacity of the sea's surfaces is produced in tension with the turbulent temporality of the wake. Offering urgency to allegory, Sharpe argues that we should "think the metaphor of the wake in the entirety of its meanings (the keeping watch with the dead, the path of a ship, a consequence of something, in the line of flight and/or sight, awakening, and consciousness)" in order to "continue to imagine new ways to live in the wake of slavery, in slavery's afterlives, to survive (and more) the afterlife of property."[17] In the wake of the slave ship, in the wake of slave transit, in the wake of slavery, the bodies of many black Africans remain, transforming the sea into a mass grave. Sharpe thinks of the wake as a rich metaphor for the material and temporal realities of slavery. Traces remain, because even for those who did not survive the holdings, or the seas, the Middle Passage or the plantation, they are alive yet, in hydrogen, in oxygen, in carbon, in phosphorus, in iron, in sodium, and in chlorine. In the chemical, material compounds of the wake, "they are with us still."[18]

Writ large against powers beyond the self, human frailty in the face of the oceanic abyss is frequently rendered as both desire and loss. In Brazilian writer Clarice Lispector's short story, "The Waters of the World," she describes a woman diving into the sea. In her words, the simple act of an early-morning swim articulates an unknowable desire between the human being and the decidedly nonhuman power of the ocean, of which we might only ever touch the smallest part: "There it is, the sea, the most unintelligible of non-human existences. And here is the woman, standing on the beach, the most unintelligible of living beings . . . She looks at the sea, that's what she can do. It is only cut off for her by the line of the horizon, that is, by her human incapacity to see the Earth's curvature."[19] The human frailty in the face of an oceanic abyss is rendered as desire across a space of human and nonhuman coexistence.

Returning to Solnit, who has written so eloquently about the politics and poetics of space, we find descriptors of the sea that articulate its power *as* metaphor. In the coda to a collection of essays on landscape and the politics of land use, Solnit is "reading the sea . . . its depths an opaque accumulation of transparencies with blue borrowed from the sky."[20] It is telling that she resorts to oceanic metaphors: "The sea always seems like a metaphor, but one that is always moving, cannot be fixed, like a heart that is like a tongue that is like a mystery that is like a story that is like a border that is like something altogether different and like everything at once."[21]

The Earth's oceans are changing, and the evidence of human intervention is prefigured in the rapid collapsing of fisheries, of glacial shelves, of coastlines. The history of oceanic representation is one deeply invested with longing, desire, and romance; with the sublime imaginings that come with the blue of distance; with the incapacity to see the Earth's curvature; with the far away and unknown; with the blood circulating within all of us. Historically, both ocean and sky have signified infinite unknowns that could be filled with fantasy projections and simultaneously with our garbage and debris. But in the contemporary moment, the metaphor is the network, and the distributed but controlled world it signifies. Material metaphors may function as abstractions of, or distractions from, environmental politics, yet they also provide methods for thinking of the ocean as a functional communications network, in contradistinction to the "corporate cloud" metaphor writ large. The ocean and the cloud have become twinned sites of elemental infrastructure, hiding digital networks that are impossible to see, while, at the same time, sea and sky are ever present in the visual culture of the twenty-first century, from the BBC's *Blue Planet II* to the popularity of drone videos on social media. Thinking through these material metaphors together—the cloud as synonymous with immaterial computing, the ocean as a networked space of globally connected cables—might yet offer new literary and aesthetic possibilities for reckoning with the Earth's largest ecosystems.

Notes

1. For an overview of the digital cloud in recent computing terminology, see Tung-Hui Hu, *A Prehistory of the Cloud* (Cambridge, MA: MIT Press, 2015).
2. Rebecca Solnit, *Field Guide to Getting Lost* (London: Penguin, 2006), 17.
3. With an estimated 100 million pieces or more of orbital debris measuring smaller than 1 cm currently in Earth's orbit, NASA recently launched a space debris sensor to monitor small debris, which is impossible to track from Earth. NASA.gov, 12 December 2017, https://www.nasa.gov/mission_pages/station/research/news/sensor_to_monitor_orbital_debris_outside_ISS.
4. In the past decade, data centers have gone from virtually nonexistent to an energy draw that consumes approximately 3 percent of global electricity supply and accounts for 2 percent of total greenhouse gas emissions. Some predictions show that the amount of energy data centers use is doubling every four years, and could triple in the next decade. See Tom Bawden, "Global Warming: Data Centres to Consume Three Times as Much Energy in Next Decade, Experts Warn," *The Independent*, 23 January 2016, http://www.independent.co.uk/environment/global-warming-data-centres-to-consume-three-times-as-much-energy-in-next-decade-experts-warn-a6830086.html.
5. Paul N. Edwards, *The Closed World: Computers and the Politics of Discourse in Cold War America* (Cambridge, MA: MIT Press, 1996), 158.
6. Plans are under way for one of the largest proposed sub-sea fiber-optic cables, which would cross the Arctic to connect Japan and Western Europe, stretching over 15,000 km. See Mark Rendell, "Competition Heats Up to Build Canada's First Arctic Fibre Line, as Inuit Want In," *CBC News*, 12 October 2016, http://www.cbc.ca/news/canada/north/inuit-fibre-arctic-project-1.3800259.
7. Nicole Starosielski, "'Warning: Do Not Dig': Negotiating the Visibility of Critical Infrastructures," *Journal of Visual Culture* 11, no. 1 (2012): 41.
8. The Porthcurno Cable Station in Cornwall, England, is an excellent example of this tendency. See http://atlantic-cable.com/CableCos/Porthcurno/.
9. Starosielski, "'Warning: Do Not Dig,'" 40.
10. Keller Easterling, *Extrastatecraft: The Power of Infrastructure Space* (London: Verso, 2014), 15.
11. Italo Calvino, "Blood, Sea," in *Textures of the Anthropocene: Grain Vapor Ray* (Cambridge, MA: MIT Press, 2014), 41.
12. This app lets users explore satellite imagery by controlling what they see through

touch-screen gestures, by "drawing" and "dragging" to form unique image combinations. See https://lines.chromeexperiments.com/.

13. Photogrammetry was understood as "the science of obtaining reliable measurements by means of photographs," and the spectacle and wonder of viewing color images was still a phenomenal experience. See Harland R. Cravat and Raymond Glaser, *Color Aerial Stereograms of Selected Coastal Areas of the United States* (Rockville, MD: U.S. Department of Commerce, National Oceanic and Atmospheric Administration, and the National Ocean Survey, 1971), 93.
14. NASA, "July 12, 1962: The Day Information Went Global," Telstar at 50, NASA, 9 July 2012, http://www.nasa.gov/topics/technology/features/telstar.html.
15. Edwards, *The Closed World*, 12.
16. Roni Horn, *Saying Water*, performance recorded at the Two Days Art-Festival, Louisiana Museum of Modern Art, Humlebæk, Denmark, May 2012, https://www.youtube.com/watch?v=fkvoe7s1NVg.
17. Christina Sharpe, *In the Wake: Blackness and Being* (Durham, NC: Duke University Press, 2016), 20.
18. Sharpe, *In the Wake*, 21.
19. Clarice Lispector, *The Complete Stories* (New York: New Directions, 2015), 401–3.
20. Rebecca Solnit, *Storming the Gates of Paradise: Landscapes for Politics* (Los Angeles: University of California Press, 2007), 382.
21. Solnit, *Storming the Gates of Paradise*, 383.

CHAPTER 4

Walking Matters: A Peripatetic Rethinking of Energy Culture

Mary Elizabeth Luka

As we humans move, work, play, and narrate with a multiplicity of beings in place, we enact historically contingent and radically distinct worlds/ontologies.

—Juanita Sundberg

The Halifax Explosion [of 6 December 1917] reverberates as a definitive historic moment around which themes of destruction, reconstruction, urbanism, and community continue to circulate. From 2014 through 2017, as the centenary of the Explosion approaches, Narratives in Space + Time Society (NiS+TS) has presented a number of public walking events designed to explore the ways in which the disaster, the ensuing relief efforts, and the reconstruction continue to shape the diverse experiences and understandings of this city.

—NiS+TS website

Responsive approaches mobilized by artists who use walking practices are crucial for reimagining worlds and refiguring our use of energy. Common to the recent turn to energy in the humanities is the recognition that energy is transferred and transformed depending on the ways it is physically and conceptually shared. I am interested in how contemporary psychogeography[1] and co-mobility[2] can be paired with the co-creative potential of creative citizenship[3] to acknowledge social histories and shift social relations in today's energy cultures. Guy Debord's theory of the *dérive* promotes the disruption of our usual patterns of social and power relationships, which restrict us from imagining new (energy) cultures, through the use of walking and mapping behaviors that challenge the status quo in the city. Furthermore, through co-mobility—that is, the experience of artists and citizens walking together in familiar and unfamiliar neighborhoods—Jen Southern illustrates how we can discover characteristics, emotions, stories, or experiences available in urban and nonurban environments that are otherwise not visible because of previously habituated uses or experiences.[4] Likewise, authors such as David Evans or Karen O'Rourke[5] provide many examples of how the experiential processes of artistic walking projects make visible the ways that power and social relations are embedded in urban, social, and, thereby, energy cultures. I am particularly interested in how co-creative practices used by media producers and artists intersect with power relations in such environments to create a shared cultural experience with narrowcast audiences.[6] This is of interest to artists and co-participants who employ walking practices as efforts to understand how the dynamics of energy culture work in relation to specific moments in time in specific communities.[7] Through the shared storytelling and sensorial walking experiences of the artist group, Narratives in Space + Time Society (NiS+TS), I elaborate how the power relations embedded in the spatial specificity of energy cultures might be reshaped and rethought. As I explain below, in the case of NiS+TS, this kind of cultural energy shift is generated and deployed through walking, by socially and digitally mediating activities, understandings, and documentation of such experiences.[8]

NiS+TS is a creative research group based in Halifax, Nova Scotia, on the east coast of Canada. It facilitates projects involving intergenerational mobile media and walking practices to elicit stories and civic engagements in often unexpected ways, connecting and disrupting history, habit, and the present. I am one of the founding members. As an artist group, we work creatively and collegially through walking, looking, talking, documenting, and sharing media. We do this *walking work* to create spaces for ameliorating traditional power relationships (class, gender, race) in the neighborhoods where we live, intervene, and participate. We tell undertold stories, make buried histories visible, and aim to balance real life (walks) with virtual experiences (our iOS app, Drifts), in order to consider and restructure specific sites of energy, power, and connectivity. By employing creative citizenship strategies and analyzing co-creative opportunities arising from our ongoing digital and mobility practices, NiS+TS is able to enact an explicitly feminist, intersectional approach to collaboration. With the help of artists, historians, architects, and communications experts, we empower and draw on the expertise of citizens on the ground.

NiS+TS was founded in late 2012, and by May 2013 it had created the "Hippodrome Project (Notes from the Desire Paths)," a workshop for seventy-five people at the defunct Montreal Hippodrome, the site of a century-old horse racetrack awaiting redevelopment as urban housing. The event was held during the international conference "Differential Mobilities," sponsored by the Pan-American Mobilities Network in collaboration with the European Cosmobilities Network. It featured food, walking activities, and the use of a mobile software application (7Scenes) to house forgotten stories, images, and sounds related to the "risky" history and potential future uses of the site.[9] The imminent redevelopment of the site itself is meant to "save energy" by intensifying the commercial and residential density of the city. But such a move also erases its colonialist history as a site of exclusion. One key lesson learned by NiS+TS was to understand the value of hosting and responsive facilitation for revealing and provocative experiences for our narrowcast audience and ourselves. Most important, the project involved accompanying groups of participants through the planned activities without leading them to predetermined conclusions. The enthusiasm of participants reshaped well-known and less-known histories of the space:

> The abandoned space and our appropriation of it temporarily, through the sharing of food, walking, conversation, and response to the physical space and to digitally-created experiences, created a new environment within which strangers to the city—scholarly, artistic, and activist visitors at the conference—could experience something other than a university classroom or a downtown restaurant . . . Multisensory, and/or mediated, the workshop was experienced (and endures in both memory and archive) as an array of possibilities and choices about whether and how to move, and alongside what narratives. In the end, the space itself completed the project. The urban ruin embraced its visitors as liminal space supercharged with history, narrative, sound, and image; always open to new interpretations.[10]

The playful nature of the experience allowed the participants to step outside their formal roles and step into the possible stories offered up by NiS+TS and its participants in these liminal moments of urban transformation. This is a much more carefully negotiated co-creative exercise than that envisioned only a decade or two ago, at the dawn of the most recent interactive age. The enthusiastic uptake in the early 2000s about the potential of the seemingly energy-efficient internet and mobile media to encourage global practices of "democratic" or "interactive" participation has since been studied only obliquely in relation to public art practice, though more widely in industrial digital media production.[11] The optimism originally expressed has become considerably tempered since 2010. Increasingly, the field reflects the challenging realities of ambitious co-creative projects that simply download work to crowds of volunteers, and then require ongoing resources to be sustainable (for example, Wikipedia's ongoing fundraising efforts and the emergence of feminist "edit-a-thons" to ameliorate gaps in putative shared knowledge). The digital humanities have articulated such concerns with citizen engagement in social relations, for example, through the Digital Humanities Manifesto 2.0 (2009). The manifesto was written by scholars and activists to assert the rights of individuals to communicate globally and freely using digital tools and approaches (for example, open-source coding), without explicitly addressing how it props up our current energy culture.[12] Its ambitious nature and scope was paralleled by the rapid development of STEM (science/technology/engineering/math) scholarship and debates through the last half of the twentieth century and particularly in the last fifteen years—including the ongoing debate about whether and how the mere insertion of "art" into STEM could create STEAM, for a new kind of energy culture.[13] The now sprawling domain of science and technology studies (STS) explicitly connects the arts and humanities to technology, and has contributed to the development of the digital humanities throughout the 2000s.

More explicitly, the notion of co-creation was reshaped by convergence culture in the 2000s.[14] Theoretical work in relation to convergence culture helpfully grapples with understanding audiences as (co-)producers, opening the door to an extension of the notion of narrowcast audiences as producers themselves, as suggested in the images (figs. 4.1–4.3) of NiS+TS audiences and their many smartphones documenting events shared with us and with others. For artistic practice, this can include specific interventions by fans or aspirational artists or groups of creative workers—including traditional producers—who influence the interpretation of cultural content, remix it, comment on it, or otherwise reshape iterations of the programming or content in question. It also includes involvement and response from programmers, curators, media, and narrowcast audiences. In creative citizenship,[15] the position of the artist or creative worker in a variety of these roles is important, as is that of a series of specific audiences, such as policy-makers, programmers, groups with explicit cultural affiliations, and so on. The collaborative nature of such media production is based on fluidly networked relationships among artists, creative producers, technical crews, programmers, broadcasters, distributors, and specific narrowcast audiences, including curators and culture supporters. But creative citizenship is also dependent on actually making or doing something. Creative citizenship can certainly incorporate the potentially generative nature of artists' work, in part because of artists' positioning as the most passionate and informed of narrowcast audiences. However, it is only through the production of creative experiences and objects themselves that the exercise of creative citizenship can be fully realized. In this context, some experiences and objects are co-created or reshaped, but not all are. Likewise, some such experiences refigure energy culture, but not all do. Understanding that creative citizenship may not be realized in every effort to refigure our energy culture reasserts the need to understand co-creation from the *professional end of the spectrum*, rather than simply from that of the audience as "produser" (producer + user).[16]

Since 2013, a suite of public art walks and concurrent digital documentation processes for exhibitions, websites, and software applications has composed NiS+TS's current overarching project, "Walking the Debris Field: Public Geographies of the Halifax Explosion." The work of NiS+TS does not just lead us toward the marking of the centenary of a traumatic, nation-building event imbricated in our contemporary energy culture. It specifically calls into question how we tell stories about colonialism, nation-state defense systems, and the particular classed and racialized groups (such as poor or working-class, Mi'kmaw, or African Nova Scotians) that were pushed out of the city center after that event. This is not simply an academic or creative exercise. It is an urgent call for optimism: to imagine and mobilize new energy cultures that can arise from artistic forms of civic engagement and build understanding about the places where we live and the people we are today.

For those who do not know the story of the Halifax Explosion, it is an exemplar of the issues related to living in an energy culture increasingly prone to violence in order to protect itself.[17] Toward the end of World War I, Halifax Harbor was filled with warships and supply ships moving to and from the European front. At night, the harbor was closed off from the ocean through the use of submarine nets, and in order to be safe, ships had to scramble to get into the harbor before night fell. On the morning of 6 December 1917, in the process of heading out of the harbor and into the open ocean, the *Imo* (a Belgian relief ship) collided with the *Mont Blanc*, which had a cargo of explosives. The *Mont Blanc* caught on fire, and twenty minutes later, it exploded, killing more than fifteen hundred people and injuring thousands of others. Hundreds of people were blinded by flying glass and debris. In total, more than 10 percent of the population of Halifax died or was critically injured. Fires broke out throughout the north end of the city and burned most of it to the ground. The event was followed the next day by a blizzard, which complicated the massive relief effort taking place. Ever since then, Halifax has held a memorial event on the morning of 6 December to commemorate those who were killed, injured, or displaced, as well as to thank those who helped the city and its occupants rebuild. While a series of stories have become embedded in the local psyche as representing the essence of the explosion and its aftermath, many of these popular or well-known stories do not provide the whole picture of the event.

The explosion operates as a traumatic event embedded in the emergence of twentieth-century energy culture in Canada.[18] The specific contours of this event are fading away and becoming lost from living memory. The waterfront, where most of the most damage occurred, is now primarily a combined military-industrial zone, while—just up the hill in the north end of Halifax—the formerly working-class neighborhood most affected by the explosion is becoming increasingly gentrified. Consequently, the emphasis at each of ten public walks, two exhibitions, and the development of an iOS app (Drifts) and augmented-reality experience by NiS+TS has been to engage specific narrowcast audiences in actively reworking narratives of the explosion. The work has included analyses of how the explosion has shaped the city, region, and country, through twentieth- and twenty-first-century preoccupations with the mediation of conflict, trauma, and recovery. The focus of each public walk emerges during research walks, where we come into contact with actual people, images, smells, and barriers. In other words, by physically, socially, and psychologically *walking through* the collective "debris fields" of such events—virtual, ephemeral, or physical—we learn, heal, adapt, and change the world around us. At NiS+TS, our most intense engagements revolve around detailed, highly personalized storytelling and sensorial experiences at sites where some of the stories not yet told (or not yet formally recognized) about the explosion originate. To elicit visceral responses, for example, during the public walks, we present expert reproductions of buildings using 3D modeling technologies to remind us of those other times and people (figs. 4.1–4.3). We then burn them, a gesture that sensorially represents the violence that often still accompanies present-day class, gender, and race relations in the city. We also source and share less familiar stories, for example, from African Nova Scotian or Mi'kmaw communities connected to specific buildings or lands affected by the explosion. Many of these stories have seldom been renarrated because these groups had (and continue to have) no recognized ownership or standing over these now highly industrialized and militarized properties. By walking through these spaces with community members, we help make visible not just how the Halifax Explosion's reverberations continue to be felt today, but also how we could imagine a different present and future.

On the public walks, storytelling, singing, image-sharing, and the burning of objects generates a moment out of time—that is, when time slips from one century to another (see fig. 4.4). Visceral acts of destruction or remembrance are preceded and followed by fictional and nonfictional explanations of historical trade routes, working conditions, and other comparisons to the present-day conditions and occupants of the specific sites we visit. Our four-year documentation of this process has resulted in sixty videos and hundreds of photos and other ephemera used online and in Drifts (the iOS app launched in October 2017). Oscillating between past and present is an evocative and powerful stimulus for participants to share their own stories about similar circumstances today, whether related to family, community, or industry. Giveaways, like scarves to cover your eyes (from our 6 December 2014 walk) or the replica mortuary bags for personal effects (from our 18 April 2015 walk), are tangible reminders to participants to give consideration to the narratives of the city and the ways in which its citizens interact (see figs. 4.5–4.6). So, too, is the creation of antimonuments, such as the commemorative concrete-and-copper *XYZ* sculpture designed and built by a dozen architecture students in the summer of 2014, and the *Psychogeographer's Table* created by another group of architecture students in the summer of 2016 and subsequently turned into both a museological artifact and an augmented-reality experience.[19] At the final NiS+TS public art walk in December 2017, almost two hundred people attended, while thousands more used the app. In other words, a community was created through engagement in shared public art experiences.

To unpack this a little more, it is worth understanding that the roots of the NiS+TS artistic practice lie in the psychogeographic practices and writing of Guy Debord in the 1950s, including the co-creative practice of the *dérive*, or "drift," and in artist-led events deriving from the Fluxus movement of the 1960s.[20] Many of us involved at NiS+TS trained or teach at Nova Scotia College of Art and Design University (NSCAD) in Halifax, an institution greatly influenced by the conceptual art movements of the 1970s. Consequently, as a group, we respond well to the giving and taking of instructions and disruptive information and activities in pursuit of quests and storytelling grounded in the spaces we occupy. We sometimes invite audiences to walk through one city, using a map from another city, just to see what that reveals. Artist projects that involve labyrinths, mirrors, tracking, or wayfinding are crucial ways to generate alternative maps of the world. The NiS+TS "Public Geographies of the Halifax Explosion" project reveals embedded, grounded power. Our projects make less abstract, and more pertinent, the deep-seated nature of capital and energy arrangements in the city, and also challenge such arrangements through co-creation, including strategies of collaboration and sensorial storytelling in an ethical manner.

Moreover, the work of NiS+TS and other artist walking projects help update the notion of the walkthrough method. Originally a set of practical

Fig. 4.1. Image from public art walk of 18 April 2015 (Aftermath). © M. E. Luka / NiS+TS.

steps for software designers and users to understand how to use and improve on software design,[21] today the walkthrough method is

> grounded in a combination of science and technology studies with cultural studies, through which researchers can perform a critical analysis of a given app. The method involves establishing an app's environment of expected use by identifying and describing its vision, operating model and modes of governance. It then deploys a walkthrough technique to systematically and forensically *step through* the various stages of app registration and entry, everyday use and discontinuation of use. The walkthrough method establishes a foundational corpus of data upon which can be built a more detailed analysis of an app's intended purpose, embedded cultural meanings and implied ideal users and uses. The walkthrough also serves as a foundation for further user-centred research that can identify how users resist these arrangements and appropriate app technology for their own purposes.[22]

Fig. 4.2. Image from public art walk of 18 April 2015 (Aftermath). © M. E. Luka / NiS+TS.

Fig. 4.3. Image from public art walk of 18 April 2015 (Aftermath). © M. E. Luka / NiS+TS.

By working together with our narrowcast audiences toward a rethinking of our shared urban spaces, NiS+TS has actively reshaped what have long been called "cognitive walkthrough methods"[23] for use in software and other design development needs. In the case of NiS+TS, we developed our design and content for Drifts, the software app, guided by several years of feedback about experiences emerging from the 1917 explosion in the Halifax Harbor, which has resonated with people in different ways and places over the past one hundred years. We did so not just to mobilize today's "walkthrough method" for virtual civic engagement processes, but also to update how "walking through" an event or historical trauma is effected in real time, in real space, and with real people. Our exploration of updated walkthrough methods allows NiS+TS to help acknowledge the urgent need to position the presence of actual humans in the obsession of today's energy culture with data collection and analysis.[24]

Although the NiS+TS engagements take place on social media and through software, they are experienced most effectively while walking through the specific locations involved. Not incidentally, such engagements provide an opening to consider landscapes as narratives and vice versa,[25] and to connect us

Fig. 4.4. Six images from public art walks of 6 December 2016 and 6 December 2017 (Turtle Grove and Centenary Procession). © M. E. Luka / NiS+TS.

to a radical rethinking of the space and resources that we humans—and our nonhuman "kin"[26]—continually consume and exhaust. To link this to the everyday, it is helpful to consider what J. K. Gibson-Graham has termed an "ethical economical and ecological engagement" to community.[27] Gibson-Graham rethinks what the economics of research and value-based social engagement could mean, and how this harkens back to a feminist understanding of multiple and fluid identities, requiring mutual commitment and care. Gibson-Graham

Fig. 4.5. Personal effects. © M. E. Luka / NiS+TS.

refers to self-positionality in the "everyday economy" as informative about long-term commitments to shared community values as well as everyday desires, pointing out that people maintain contingent and coexisting "identities in excess of . . . those associated with capitalist practice."[28] Such an approach points toward "the possibility of [employing] ethical rather than structural dynamics" for reshaping our current energy culture in as-yet-unimagined ways.[29]

The content-sharing, genre-expanding, and narrowcast audience projects with which NiS+TS is involved reflect its messy, conflicted, and effective spacemaking efforts as a co-creative undertaking by a team of creative facilitators, collaborators, and an increasing number of narrowcast audience members. NiS+TS public art walks include intense attention paid to the sensory, to shared stories, and to narrowcast audience–generated media uploaded to social media platforms. From this viewpoint, it becomes possible to map not simply what is familiar about energy cultures but also what is unnoticed about the terrain we travel every day in relation to energy culture—and, further, to understand energy *as culture*. Well-established methods such as Debord's *dérive*[30] can help to unsettle the constraints and habits of thinking developed

Fig. 4.6. Antimonument. © M. E. Luka / NiS+TS.

over time in urban and nonurban environments, as can specific technological and community-building tools used by NiS+TS and other artists. The approach aligns itself with investigations of digitally powered "augmented reality"[31] to suggest ways to imagine life beyond "petrocultures."[32] In other words, it is by crowding together, mobilizing in neighborhoods, and listening, recording, editing, and sharing that we really move toward co-creation and toward the co-imagining of acts of acknowledgment, inclusion, and engagement with new potential energy cultures.

Notes

1. Guy Debord, "Theory of the Dérive," in *Situationist International Anthology*, translated and edited by Ken Knabb (Oakland, CA: Bureau of Public Secrets, 2006 [1958]); Tina Richardson and Ally Standing, eds., *Stepz II: Between the Rollerama and the Junkyard* (Manchester: People's History Museum, 2016).
2. Jen Southern, "Comobility: How Proximity and Distance Travel Together in Locative Media," *Canadian Journal of Communication* 37, no. 1 (2012): 75–91.

3. Mary Elizabeth Luka, "This Is Not a Gift Shop: Arts-Based Narratives of the Canadian North," *Topia: Canadian Journal of Cultural Studies* 32 (2014): 111–34.
4. Southern, "Comobility."
5. David Evans, ed., *The Art of Walking: A Field Guide* (London: Black Dog Publishing, 2012); Karen O'Rourke, *Walking and Mapping: Artists as Cartographers* (Cambridge, MA: MIT Press, 2013).
6. Luka, "This Is Not a Gift Shop."
7. O'Rourke, *Walking and Mapping*; Juanita Sundberg, "Decolonizing Posthumanist Geographies," *Cultural Geographies* 21, no. 1 (2014): 33–47.
8. Robert Bean, Léola Le Blanc, Brian Lilley, Barbara Lounder, and Mary Elizabeth Luka, "Narratives in Space and Time Society (NiS+TS): The Hippodrome Project," in *Urban Encounters: Art and the Public*, edited by Martha Radice and Alexandrine Boudreault-Fournier (Montreal: McGill-Queen's University Press, 2017).
9. Bean, Le Blanc, Lilley, Lounder, and Luka, "Narratives."
10. Bean, Le Blanc, Lilley, Lounder, and Luka, "Narratives," 110–11, 125.
11. Henry Jenkins, Sam Ford, and Joshua Green, *Spreadable Media: Creating Value and Meaning in a Networked Culture* (New York: New York University Press, 2013); Douglas Rushkoff, *Open Source Democracy: How Online Communication Is Changing Offline Politics* (London: Demos, 2003); Rushkoff, *Present Shock: When Everything Happens Now* (New York: Penguin, 2013); Don Tapscott and Anthony D. Williams, *Wikinomics: How Mass Collaboration Changes Everything* (New York: Portfolio, 2008); and Liesbet van Zoonen, *Entertaining the Citizen: When Politics and Popular Culture Converge* (London: Rowman & Littlefield, 2005).
12. "The Digital Humanities Manifesto 2.0," A Digital Humanities Manifesto, 29 May 2009, http://manifesto.humanities.ucla.edu/2009/05/29/the-digital-humanities-manifesto-20/.
13. See, for example, *Transformations* 26, "Thinking in the Arts-Science Nexus."
14. For studies of fan culture, see Henry Jenkins, *Fans, Bloggers, and Gamers: Exploring Participatory Culture* (New York: New York University Press, 2006). On democracy, see Rushkoff, *Open Source Democracy;* and Rushkoff, *Present Shock*. On Wikipedia and crowdsourcing, see Tapscott and Williams, *Wikinomics*. On producer/consumer relations, see William Uricchio, "Beyond the Great Divide: Collaborative Networks and the Challenge to the Dominant Conceptions of Creative Industries," *International Journal of Cultural Studies* 7, no. 1 (2004): 79–90.
15. Luka, "This Is Not a Gift Shop"; and Mary Elizabeth Luka, "CBC ArtSpots and the Activation of Creative Citizenship," in *Production Studies, Volume II*, edited by Vicki Mayer, Miranda Banks, and Bridget Conor (New York: Routledge, 2016).
16. Rebecca Ann Lind, ed., *Producing Theory in a Digital World: The Intersection of Audiences and Production in Contemporary Theory* (New York: Peter Lang, 2012).
17. For a descriptive, mixed-media, web-based project developed by the Canadian public broadcaster with the help of community experts about fifteen years ago, see http://www.cbc.ca/halifaxexplosion/. See also Janet Kitz, *Shattered City: The Halifax Explosion and the Road to Recovery* (Halifax: Nimbus Publishing, 2008); and Janet Maybee, *Aftershock: The Halifax Explosion and the Persecution of Pilot Francis Mackey* (Halifax: Nimbus Publishing, 2016), for examples of local historical projects that aim to "set the record straight."
18. Ruth Sandwell, ed., *Powering Up Canada: The History of Power, Fuel, and Energy from 1600* (Montreal: McGill-Queen's University Press, 2016).

19. The *Psychogeographer's Table* was one of several elements presented during an exhibition at the Dalhousie Art Gallery from October to December 2017 (http://artgallery.dal.ca/halifax-explosion), and was subsequently installed for research and exhibition through 2018 at the Maritime Museum of the Atlantic's centenary exhibition titled "Collision in the Narrows" (https://maritimemuseum.novascotia.ca/what-see-do/special-2017-exhibits).
20. O'Rourke, *Walking and Mapping*, 7, 12–13, 156.
21. Cathleen Wharton, John Rieman, Clayton Lewis, and Peter Polson, *The Cognitive Walkthrough Method: A Practitioner's Guide* (Boulder: Institute of Cognitive Science, University of Colorado at Boulder, 1994).
22. Ben Light, Jean Burgess, and Stephanie Dugay, "The Walkthrough Method: An Approach to the Study of Apps," *New Media & Society* 1, no. 20 (2016): 811 (my emphasis).
23. Wharton, Rieman, Lewis, and Polson, *The Cognitive Walkthrough Method.*
24. Jacob Metcalf and Kate Crawford, "Where Are Human Subjects in Big Data Research? The Emerging Ethics Divide," *Big Data & Society* 3, no. 1 (2016): 1–14.
25. Katrín Lund, "Landscapes and Narratives: Compositions and the Walking Body," *Landscape Research* 37, no. 2 (2012): 225–37.
26. Donna Haraway, "Anthropocene, Capitalocene, Plantationocene, Chthulucene: Making Kin," *Environmental Humanities* 6, no. 1 (2015): 159–65.
27. J. K. Gibson-Graham, "Ethical Economic and Ecological Engagements in Real(ity) Time: Experiments with Living Differently in the Anthropocene," *Conjunctions: Transdisciplinary Journal of Cultural Participation* 2, no. 1 (2015): 47.
28. Gibson-Graham, "Ethical Economic and Ecological Engagements," 56.
29. Gibson-Graham, "Ethical Economic and Ecological Engagements," 57.
30. Debord, "Theory of the Dérive."
31. Mark Graham, Matthew Zook, and Andrew Boulton, "Augmented Reality in Urban Places: Contested Content and the Duplicity of Code Augmented Realities," *Transactions of the Institute of British Geographers / Royal Geographical Society* 38, no. 3 (2013): 464–79.
32. Sheena Wilson, Adam Carlson, and Imre Szeman, eds., *Petrocultures: Oil, Energy, Culture* (Montreal: McGill-Queen's University Press, 2017).

CHAPTER 5

Several Documents Pertaining to the Cascade Energy (transition) Park Corporation Corporation (CORPCORP)

Marissa Benedict, Cameron Hu, Christopher Malcolm, and David Rueter

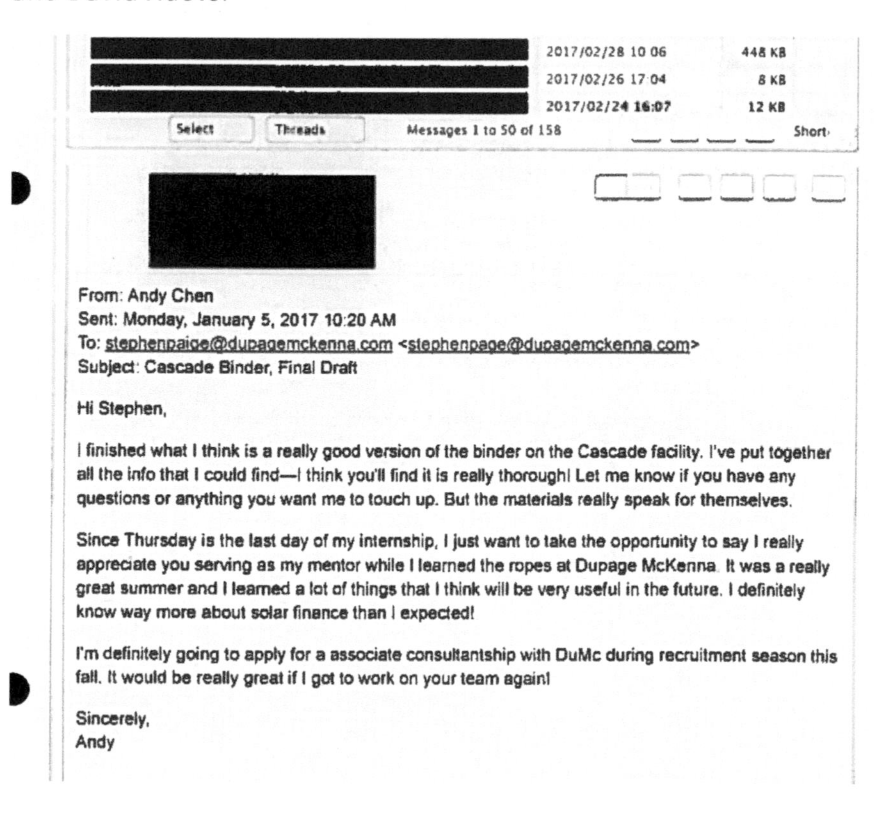

[redacted]	2017/02/28 10 06	448 KB
[redacted]	2017/02/26 17 04	8 KB
[redacted]	2017/02/24 16:07	12 KB

Select Threads Messages 1 to 50 of 158 Short

From: Andy Chen
Sent: Monday, January 5, 2017 10:20 AM
To: stephenpaige@dupagemckenna.com <stephenpage@dupagemckenna.com>
Subject: Cascade Binder, Final Draft

Hi Stephen,

I finished what I think is a really good version of the binder on the Cascade facility. I've put together all the info that I could find—I think you'll find it is really thorough! Let me know if you have any questions or anything you want me to touch up. But the materials really speak for themselves.

Since Thursday is the last day of my internship, I just want to take the opportunity to say I really appreciate you serving as my mentor while I learned the ropes at Dupage McKenna. It was a really great summer and I learned a lot of things that I think will be very useful in the future. I definitely know way more about solar finance than I expected!

I'm definitely going to apply for a associate consultantship with DuMc during recruitment season this fall. It would be really great if I got to work on your team again!

Sincerely,
Andy

426

A Partial Timeline of the Cascade Energy Park

1989 Nestle sells its NEMC (Nestle Electronic Materials Company) sub-unit, including semiconductor research and production facilities in Aceh and West Kalimantan, to the German public limited company Grube AG.

1995 NEMC stock begins trading on the NYSE with an initial public offering of $600 million.

2003 A US-led coalition launches Operation Iraqi Freedom.

2005 NEMC announces its acquisition of ComEnergy, North America's second-largest solar energy services provider. CEO Eric Ma announces that NEMC "will now participate in the development of advanced power facilities and commercialization of future-oriented energy."

2006 NEMC closes its semiconductor facilities in Southeast Asia and changes its name to ComEnergy.

2008 West Texas Intermediate crude oil prices reach $147 per barrel.

2011 Axio Power commences construction of its Cascade Energy Park in San Bernardino County, CA.

2012 NEMC acquires Axio Power and its facilities, among them the Cascade Energy Park. Brookfield Asset Management establishes the first YieldCo, "Brookfield Renewable Energy Partners."

2013 Green Energy Investor reports: "Yieldcos are a win-win product for both the investors and the operators. They lower the cost of capital which makes solar energy cheaper, which will lead to more farms and more yieldcos. Investors not only get a higher yield compared to normal gas or coal assets, but they also get to invest in a green venture."

2014 NEMC transfers the Park to its newly-established YieldCo subsidiary EarthSurge Power, raising $675 million dollars in an initial public offering. EarthSurge Power, whose website lists a PO Box in Bethesda, Maryland as its primary address, has no employees.

2015 ComEnergy and 32 'affiliated debtors' file voluntary petitions for relief under Chapter 11 of the United States Bankruptcy Code in the United States Bankruptcy Court for the Southern District of New York.

Filename: 2016-07-15_S2_D_B_video transcription.txt
Length: 1 of 2 pages

[video transcription] S2█, P. D█, and M. B█ stand on a single paved road between two project phases. The road was the only one of its type, made possible, they later discovered, by a successful application for a Major Variance at the project's inception. They were otherwise surrounded by dirt paths and desert.

Document type: video still
Filename: 2016-02-08_vicinity.mov
Length: 32:00 min

"In its initial stage, of course," said B█, "the project was encouraged by the county's General Plan and promised to have highly beneficial environmental impacts." B█ holds a partial copy of the plan and leafs through its pages.

D█: "Remember, EISs and EAs are in essence attempts to predict the future. By its very nature, predicting the future is difficult. Documents of this kind should acknowledge that difficulty and then clearly set out how the Department should proceed."

D█: "All land use requires a set of coherent development policies. Land use is the most visible expression of our work. All development, even where there is nothing, bears the mark of our decisions. Because it governs how land is to be utilized, virtually all of the issues and politics contained in other elements relate in some degree to the Land Use Element."

B█: "This must have been zoned originally as RL, and then either changed to RC, or been RC all along. Pointing toward the mountain, "the scenic vista from that range would have been disturbed. Residents would no doubt have objected, and an objection on aesthetic grounds must also be considered, but they understand that growth is inevitable. It's clearly a code three in any case "Less than Significant Impact with Mitigation Incorporated." In which case, although the proposed project could have a significant effect on the environment, there shall not be a significant effect in

Filename: 2016-07-15_S2_D_B_video transcription.txt
Length: 1-2 of 2 pages

WERE ADDITIONAL VIABILITY STUDIES CONDUCTED?

this case because revisions in the project have been made by or agreed to by the project proponent. A MITIGATED DECLARATION would have been be prepared."

B■■■: "Please recall, the details of a) Less than Significant Impact. The General Plan Open Space Element Policy OS 5.1. states that a feature or vista can be considered scenic if it: Provides a vista of undisturbed natural areas; Includes a unique or unusual feature that comprises an important or dominant portion of the viewshed; or, offers a distant vista that provides relief from less attractive views of nearby features (such as views of mountain backdrops from urban areas). The site is not part of a vista of natural areas, as surrounding areas are generally flat and intervening landscapes and man-made structures limit views. More distant vistas from higher-elevation areas are not significantly impacted due to the low height of the proposed solar panels and other project features."

[B■■■ in conclusion] "There are no unique or unusual features on the site that could dominate views of the area. Therefore, there are no unique or unusual features on the site that could comprise an important or dominant position in the viewshed."

D■■■■: "you are right of course, B■■■. D■■■■: "our work is nothing but our work, it puts a distinction where there was not one previously; the common is no use unless it can be appropriated. Indeed, there is no idea of the common without an idea of use. The land is indistinct until it is subdued, tilled, sowed and thereby annexed---[S2■■■ barely audible]: "I long for scenes where man hath never trod."--[Pointing to the BLM land to the North-East, D■■■■ continued] "in leaving as much as another can make of, we do as good as take nothing at all. Nobody could think himself injured by the drinking of another."

Document type: video still
Filename: 2016-02-08_vicinity.mov
Length: 32.00 min

Filename: 2016-07-15_S2_D_B_video transcription.txt
Length: 2 of 2 pages

B■■: "Every project is environmentally just, otherwise it wouldn't proceed." B■■ verbatim "The fair treatment and meaningful involvement of all people regardless of race, color, national origin, or income with respect to the development, implementation, and enforcement of environmental laws, regulations, and policies is assured. Fair treatment means that no group of people, including racial, ethnic, or socioeconomic group should bear a disproportionate share of the negative environmental consequences resulting from industrial, municipal, and commercial operations or the execution of federal, state, local, and tribal programs and policies. We identify potential indicators, we establish potential effects to those indicators, then we determine whether those indicators are to be significantly affected. If an indicator is to be affected significantly, that is, if within the spatial and temporal boundaries established, and with due attention given to the magnitude and duration of the impact in question it has a right of claim, mitigation measures are applied to resolve the issue. After all, Mitigation involves taking steps to avoid, minimize, rectify, reduce, eliminate, or compensate for the impact of an analyzed alternative. If the issue remains unresolved then the effect is named as residual and is to be analyzed as part of a Cumulative Effect."

western portion of this waste is called the Mojave Desert—and I here use the word waste advisedly, for desert is a misnomer when applied to a region more like that of the Mosaic myth. The boundaries of that particular part of California known as the Mojave Desert are not well defined, especially on the southward and eastward sides, on which it merges imperceptibly into the Colorado Desert; but it may be considered as occupying the north-western portion of San Bernardino County, the south-eastern corner of Kern County, and the north-eastern corner of Los Angeles County, the greater part of its area lying within the first-named county.

The general surface of the desert is from two thousand to two thousand five hundred feet above the level of the sea, and is curiously broken by ranges of hills, often rising to the dignity of mountains, and capriciously running in every direction, as well as by detached hills, or 'buttes'; while there level portions of the plain are traversed by ridges of sand, not unlike the dunes of the seaside, and like them, the effect of the high winds which sweep over the loose soil. At times, especially during the so-called spring and autumn of southern California, this sand, and pebbles even larger than a hazel-nut, are blown hither and thither, like drifting snow. In the gullies on the northern and western flanks of the hill, this material has accumulated during the ages to an extraordinary height above the desert level, and up the slopes the sand-loving yucca and muscru climb. No bowlders are to be found on the desert, but angular fragments and water-worn pebbles of quartz, flint, and igneous rocks are strewn everywhere.

MORE CURRENT ENVIRO SURVEY OF THE SITE?

The air of the desert is so dry and pure, that distances are exaggerated and objects magnified. The mirage simulates water on the surface of the dry lakes: to the eastward in the morning, to the westward when the sun has passed the meridian. Seen in the exquisitely clear atmosphere attending the dawn of a midsummer day, the distant mountains are of a deep indigo blue, which becomes lighter as the sun mounts up above them; and at this hour, while the western sky is amethystine with reflections from the yet un-risen sun god, the whole eastern rim of the desert dances in fantastic mirage, resembling, in effect, the weird tumbling of the surges, when one at sunrise looks thitherward from the deck of a ship in mid-Atlantic. In the glare of the fierce sun of noontide the hills appear in their true light—cold, gray, and desolate, each jagged peak distinct and clearly cut against a sky as cloudless as that of Italy.

WE ARE ON A SIMILAR VOYAGE

[S2█ seems to be roaming the area (audible footsteps); audio appears to have recorded directly into the survey app] "the site slopes slightly downward at an approximately 1 percent gradient towards the east and northeast. Elevations range from 2,370 feet at the northeast corner of the site to 2,400 feet at the southwest corner. Project soils are quaternary alluvium deposits (Qal), consisting of loose to medium-dense sands underlain (at depths of 10 to 15 feet) by complex mixtures of fine sand, silt, and clay. At first sight, plant communities in the area include Mojave creosote bush (larrea tridentata) scrub, allscale scrub, white bursage scrub, and big galleta grassland. Creosote ring, as yet unsubstantiated."

[S2█ continued] "Creosote ring is believed to be one of the first life forms to colonize the Mojave Desert when the last glacier receded, and has been a continuous resident here ever since. Estimates put some varieties at over 11,000 years old."

"The creosote rings" [S2█ elaborating without prompting] "are actually clone assemblages, a phenomenon in which a parent plant undergoes, over a long period of time, various stages of clonal expansion with enormous vegetative persistence resulting in living things that may surpass even bristlecone pines in longevity. Vasek was the first to see a ring of separate creosote bushes as all part of the same living thing. He theorized that the entire distance to the center of the creosote ring was at one time solid wood, the outer ring of bushes comparable with the outer layer of living bark on a redwood tree. But proving his theory became a detective story, because the heartwood inside wood and therefore the growth record, had long since rotted away. Yet Vasek was required to determine if all the bushes in a ring were genetically the same. If they were, the age of that single individual could be estimated by the rate of growth of the plant outward from the center. Of course, the original stems are no longer present, and the rings of vegetation, called circular clones, are empty inside. Vasek told me he determined the age by extrapolating present growth rates back into their absent past. Creosote rings are old, but their history cannot be read."

GENERATE FOLLOW-UP REPORT – SEND TO J IN RESEARCH

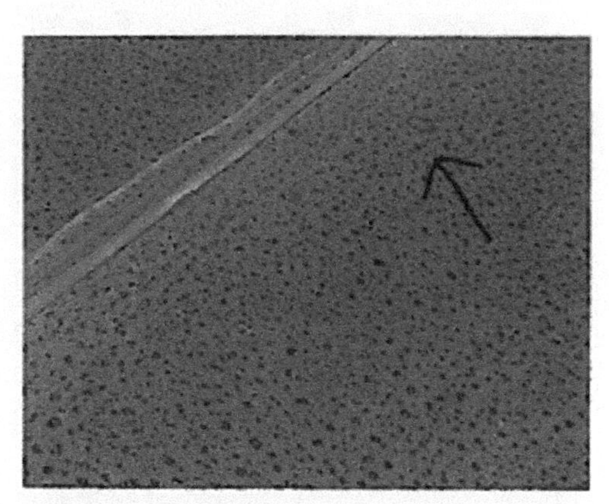

Document type:	image
Filename:	KingCloneAerialView.jpg
Length:	N/A

Perimeter Fencing

Eight-foot-tall fencing is proposed along the perimeter of the project site. Fencing would consist of seven feet of chain link topped with approximately one foot of three-strand barbed wire. Access gates would be provided at four locations (two locations on Broadway and one each at Sunflower Road and 4th Street). Security cameras would be placed at appropriate locations to deter potential trespassers.

FULL BRIEF ON SITE SECURITY AND RISK ASSESSMENT?

Exploration of the circumference of the perimeter and attendant security systems
01/03/2017

3:00 PM. In the daylight, it became clear that the perimeter did not employ a capacitive proximity sensor, but instead a fence-mounted fiber-optic style breach/vibration detection system (or perhaps a microphonic wire—the two can look similar). The system thus did not communicate a concern about potential intruders surveying the grounds at a remove, only actual breaches.

3:05 PM. Proceeded to walk counterclockwise around the perimeter, following the fenceline. Proximity and pattern seem to indicate that the downed towers noted earlier must have powered the gate entry systems. No camera systems observed.

3:35 PM. Roughly 3km past the downed towers, a second fence of similar construction started just 1m outside the perimeter fence, forming a narrow corridor. Proceeded into the corridor, as no gate or visible surveillance system hinted that access was restricted or monitored. It was unclear, however, why the new fence also had a breach detection system, as it did not appear to enclose anything.

CONSIDER REMOVAL – SEND TO D IN SYSTEMS ANALYSIS

4:15 PM. Walked nearly 5km through the corridor, through several sharp right-angle turns. Stopped to drink some water, eat an energy bar, and check for a mobile signal (nothing). The corridor showed no sign of ending. The sun was starting to set, but today was the last chance to explore the park, so turning back wasn't an option.

5:05 PM. Instead of continuing around a corner, the corridor came to an abrupt end, nearly 8 km from the entrance. A lone solar panel atop a pole at the intersection of the fences powered the breach detection system connected to the exterior fence. A microwave proximity sensor, mounted on the pole pointed directly at the corridor, prompted a momentary fear of detection and recognition of the impossibility of escape. Closer inspection revealed that the unit was not attached to any power source or communication line, and was perhaps only of symbolic value.

Filename: _DSC2136.jpg

Filename: _DSC2168.jpg

Filename: _DSC2260.jpg

Filename: DSC2246.jpg

12/30/2017 J IN MARKETING OR MANAGEMENT? OR RESEARCH?

At J.'s bungalow in the mountains, out past the military base. In the last e-mail J. said he'd arrive by breakfast, and then we would head out together in his jeep and begin work on the survey (now well past deadline). It's evening now, and still no sign of him.

By dusk I wrote the day off and set out for a walk. Headed east out of the bungalow, the sun mostly obscured behind the mountains to my back. After two short blocks the paved road gave out, thereafter continued onto a dirt path. Some ten minutes later the large installation, on which the following paragraphs will provide a cursory report, came gradually into view. In the remains of daylight its size was difficult to estimate—maybe three quarters of mile on the N/S dimension.

Another fifteen minutes to the edge of the installation, which now spread out in either direction to a vanishing point. Pointed the electric torch at the perimeter fence, illuminating a grey cable affixed horizontally to the chain-link mesh, running just beneath waist-height. Think the cable could power a capacitive sensor array or a breach detection circuit, connecting nodes in the installation's security system. Thereafter, kept more distance from the perimeter, worried about an alarm. In the distance a red and blue police beacon flashed continuously, but did not see it change position.

Followed the fence north for several hundred feet, the installation on my right, up to a newly-paved road. Here the enclosure broke, with a second fenced-off compound sprawling northward on the opposite side of the road. Turned east, into the channel between the enclosures, and resumed walking.

At some distance several lights could be seen seen bobbing gently—the headlights of four-wheelers carving down the hill. The lights grew larger and for a few minutes they seemed headed straight in my direction. Then, somewhere to the right, just beyond the southern edge of the installation, an animal, likely a goat, bleated all of a sudden and loudly—indeed, am now reminded that the whole place smelled of manure—and on turning back to the road the lights had curved southward and begun to disappear, one after another, into the night.

Circled the northern enclosure and made haste back to the bungalow. In total, the trip lasted two hours. Also seen: a group of campers pitching tents, illuminated by the hi-beams of a new hybrid.

Filename: 2016-12-15_notes_video transcription.txt
Length: 1 of 1 page

[Video transcription]

[nightfall]

[S2■■ focusses on a small grain of hi-desert sand.] "Much more granular than the Sonoran or Colorado, the elevation in the Mojave can be seen in its sand." [Looking up at the sky S2■■ seems to find an association of the small rocky substance in hand with the stars above. S2■■ is moved to whisper] "I think and speak of other things / To keep my mind at rest. The night wind whispers in my ear, The moon shines in my face; A burden still of chilling fear I find in every place..." [breaking off, S2■■ wonders aloud] "Is the sky conceived as a roof that shelters us by anchoring us in the world, standing on a horizontal plane, under the sky, reassuringly stabilized by the weight of our own gravity? Or is the sky, not of earth at all? We start to fall, so to speak, skyward, away from gravity. If the sky merely appears, as in our inability to understand it fully, then it is no shelter."

[D■■■ adjacent]: "In harnessing the Sun, do we appropriate it?" [For days, D■■■ has been chewing over something that the Interior Secretary had said in a recent statement to the press] "we are here, here, where the source of the energy never goes away." [D■■■ opens a laptop. Almost by accident D■■■ had created a 230 page document, titled, "The Liability of the Sun," 47 pages of which were devoted to an ill-tempered section in which D■■■ mused on the puzzle] "given that I never chose to revolve around the Sun, what are the reliable alternatives?" CAN WE LOCATE TH[...] DOC[...] FOR

[B■■ in the vicinity of S2■■ and D■■■] "Is it not simply acceptable that this is occurring? Has it not already occurred and given that it is here then by right is it not acceptable as such? Why am I here?" [B■■ admonishes themselves] "Do not characterize impacts as "acceptable." Use quantitative comparisons or words such as "very small." [B■■ continued] "do not subtly play down alternatives that DOE does not prefer. Provide professional, authoritative, and dispassionate responses. And even though the decision is sure, use conditional wording and verb tense, such as "would" rather than "will". Use maps and drawings to depict all features that are needed to understand the project and its impacts; provide directional arrows and scale indicators. Use existing authoritative definitions as much as possible: EXAMPLE: The Baseline Case. The baseline case provides a standard reference for assessment, which describes the environmental conditions in the relevant LSA and RSA prior to development. The Baseline Case considers existing and approved projects and activities (i.e., approved by federal, provincial or municipal regulatory authorities) as a standard and expectation."

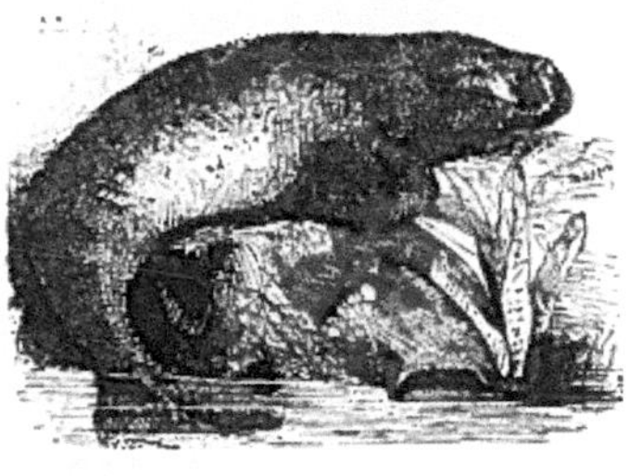

Document type: image scan
Filename: S2_sketch_logbook.jpg
Length: N/A

[S2■■ wandering into the distance] "We have lost from our collective memory any notion of the former abundance of plant and animal populations. Whales, manatees, dugongs, sea cows, monk seals, crocodiles, codfish, jewfish, swordfish, sharks and rays are large marine vertebrates, for instance, that are now functionally or entirely extinct in most coastal ecosystems."

Filename: 2016-12-15_notes_video transcription.txt
Length: 1 of 1 page

There are dozens of places in the Carribbean named after large sea turtles whose adult populations now number in the tens of thousands rather than the tens of millions of a few centuries ago. For our ancestors, abundance was perceptual. For us, disappearance is.*

—REMEMBER TO CALL D.
BEFORE GULF SITE VISIT

THE WALL STREET JOURNAL.

BLOGS

Home World U.S. New York Business Markets Tech Personal Finance Life & Style Opinion Careers Real Estate Small Business

Tech

WSJ on the cases, trends and personalities of interest to the business community

JANUARY 3, 2017, 5:20 PM ET

ComEnergy Files for Chapter 11, Raising Questions about Viability of YieldCos

Article Comments (18) TECH HOME PAGE

By

Energy giant ComEnergy filed for bankruptcy protection on Tuesday, pledging to curb a debt-fueled global expansion that pushed the company's stock to great heights before initiating its rapid collapse.

The filing caps a dramatic decline for a company that was worth over $10 billion six months ago, when it was hyped to become a global clean-energy giant. ComEnergy used a combination of financial engineering and cheap debt to establish renewable-power projects around the world before the market turned sour last summer amidst low oil prices and rising interest rates.

ComEnergy CEO Eric Ma stated the company would use the bankruptcy process to reduce its borrowings which stand at more than $17 billion, including the debt of two publicly-traded subsidiaries EarthSurge Power and EarthSurge Global Inc. Those subsidiaries—separate entities known as "YieldCos" that buy operating projects from ComEnergy and pay out cash flow to their shareholders.

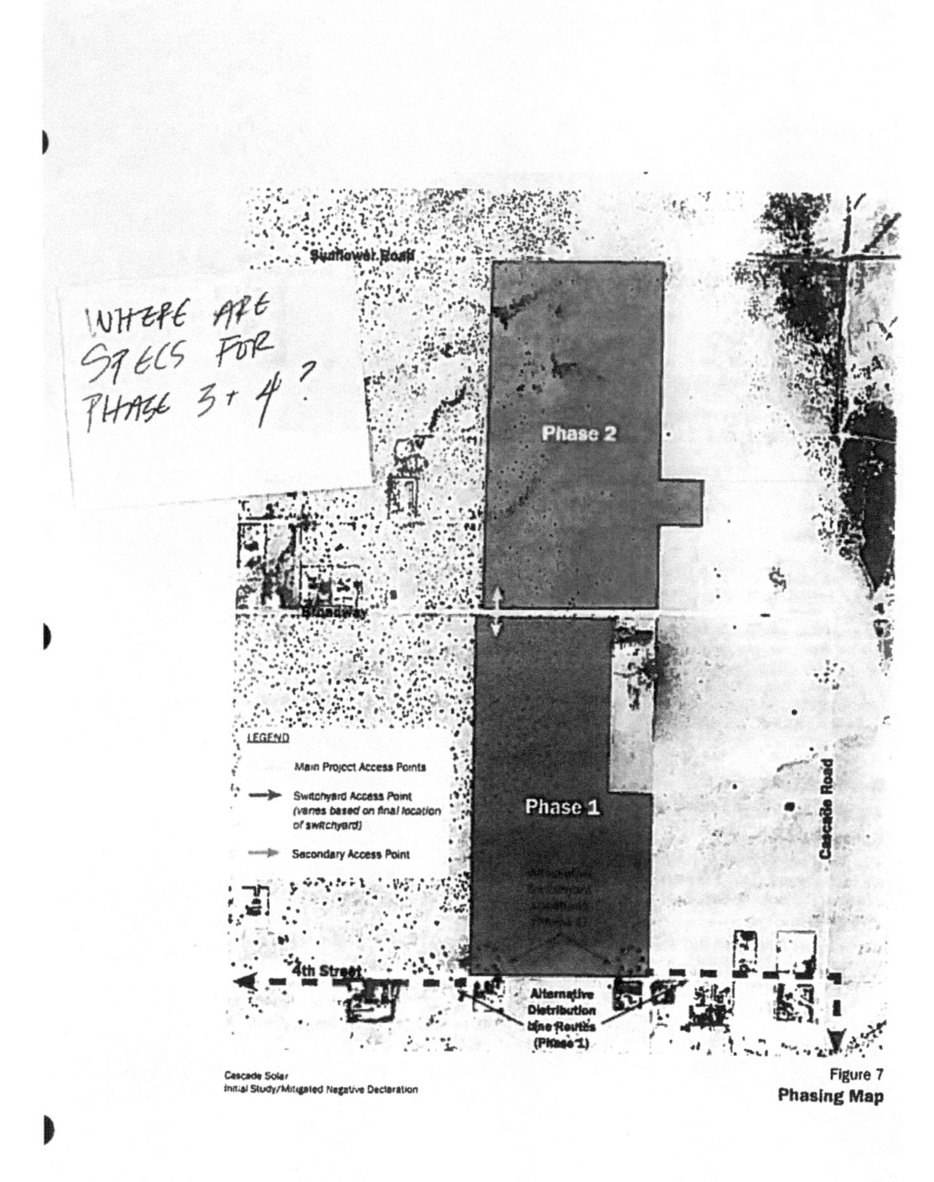
WHERE ARE SPECS FOR PHASE 3 + 4?
Phase 2
Phase 1
LEGEND
Main Project Access Points
Switchyard Access Point (varies based on final location of switchyard)
Secondary Access Point
Cascade Road
4th Street
Alternative Distribution Line Routes (Phase 1)
Cascade Solar
Initial Study/Mitigated Negative Declaration
Figure 7
Phasing Map

CHAPTER 6

Sustaining Petrocultures: On the Politics and Aesthetics of Oil Sands Reclamation

Jordan Kinder

We will ensure the land disturbed by our operation is returned to a stable, safe condition that is capable of supporting biologically self-sustaining communities of plants and animals.

—Syncrude Canada

Many of the human-altered landscapes of the present appear to be landscapes beyond resurrection.

—Alberto Toscano, "The World Is Already without Us"

In a short promotional video of 2011 issued by the Canadian Association of Petroleum Producers (CAPP), a man in a hard hat, high-visibility safety vest, and sunglasses strolls through a lush landscape.[1] Uplifting piano music plays in the background. He opens his hands, gesturing toward the landscape that surrounds him. "This is an active mining operation in the Canadian oil sands," he says. "It's not a pretty sight when you open up the Earth in order to extract the oil, but after this operation is finished, it will be reclaimed." And at the end, "Where there was once an oil sands mining operation, you now have a beautiful biodiverse landscape again," the narrator concludes. "You'd never know there'd been a mine there in the first place." A white screen with black text fades in, declaring: "New ideas are making a difference." The landscape documented in this video is a reclamation project in Alberta's oil sands, a greenspace nested within a wider landscape of active mines, seemingly endless deforestation, and general destruction—a "pretty sight" amid a not-so-pretty one. This video features Patrick Moore, co-founder of Greenpeace and self-described "sensible" environmentalist.[2] His endorsement of the project strategically deploys his environmentalist cultural capital to imbue reclamation—and bitumen extraction in general—with "green," renewable characteristics. As is well known,[3] however, the reclamation process, like the process of mining and refining bitumen that necessitates such projects in the first place, is extremely resource intensive and has historically been developed in a touch-and-go fashion, only to be later discursively repackaged retroactively as "innovation"—as "new thinking."[4] Such a description of reclamation—as innovation, as new thinking—strategically avoids the realities of the intensities and uncertainties of reclamation. When considered this way, the whole project of oil sands reclamation works to superimpose the limitless logic of capital onto the limits of the nonhuman world, and it is within this disjuncture in possibilities for reclamation that I focus the present intervention.

All land rented by oil companies in Alberta must by contract be "reclaimed" at some end point, and that reclamation clause specifies that the land is returned "to an equivalent land capability." "If an area meets stringent requirements for reclamation," the government of Alberta states, "regulators will issue final certification and the land is returned to the Crown as public land. To date, one area called Gateway Hill is certified reclaimed."[5] Syncrude's Gateway Hill covers 104 hectares, nestled in 89,592 hectares of total land affected by oil sands mining.[6] Distinctions between types of landscapes and their current possibility for reclamation are worth pointing out here, especially in the terms of the distinction between agricultural landscapes, which have been "human-altered" for generations, and what are considered wild or natural landscapes. There is little evidence that wild or natural landscapes can be reclaimed, even according to the vague metric of "equivalent land capability," despite the legal requirement for companies operating in the oil sands to commit to this process.

Outside of the hard sciences, little research exists on oil sands reclamation projects.[7] This gap in critical attention is significant because reclamation projects such as Gateway Hill represent a unique site from which to develop a comprehensive, micro- and macro-critique of petrocapitalism, with particular attention to the ways it engages and affects the human and nonhuman elements involved. Gateway Hill offers a site at which to explore and interrogate the relationship between bitumen extraction, oil sands reclamation, and the kinds of techno-scientific processes that shape contemporary discourses and cultural imaginaries surrounding oil extraction and other climate-altering processes. Rather than climate or ecological "remediation," the techno-utopian processes embodied in reclamation are instead a kind of *capital* remediation, an attempt to mitigate the ever intensifying contradictions of petroculture and petrocapital, which always already falls short in its promises by failing to adequately recognize that energy is social and cultural as much as it is physical. I approach reclamation through three interrelated vectors and their relation to what is at stake, culturally and materially, in the claim of the possibility of returning a landscape to "equivalent land capability": human and nonhuman production in the productivist imaginary, the aesthetics of oil sands reclamation, and oil sands reclamation's techno-scientific impetus.

Reclamation's Materials: The Production of Nature and the Nature of Production

At the core of the reclamation claims of the possibility of returning a damaged landscape to "equivalent land capability" are questions of use-value and exchange-value, on the one hand, and nature and production, on the other. For Marx, the concept of nature is deeply rooted in his understanding of production, the creation of value, and of use-value in particular, which can be summarized in the following formula: labor plus nature equals production. A tempting impulse emerges here to critique Marx's calculus as a perpetuation and reproduction of a kind of binary, Enlightenment view of the oppositional relationship between the human and nonhuman, and society and nature. But, for Marx, this is not the case. As Alfred Schmidt, theorist of Marx's ecological thought, notes, "Marx considered nature to be 'the primary source of all instruments and objects of labour,' i.e. he saw nature from the beginning in relation to human activity."[8] Schmidt summarizes: "Nature was for Marx both an element of human practice and the totality of everything that exists."[9] Use and instrumentalization, a notion tied to a productivist imaginary,[10] is complicated here in the way that Marx views nature as an aspect of "human practice," as well as a totality comprising everyone and everything that exists. Not so much an attempt to perceive and characterize nature as a space for extraction, that is, for purely *human* use, Marx's view of nature arguably opens up several ways of accounting for the complex relationship between humans and nonhuman nature. This dynamic is embodied in Marx's development of the notion of the metabolic rift, a concept that ecosocialists such as Paul Burkett and John Bellamy Foster revive in relation to contemporary ecological relations under capitalism.[11] McKenzie Wark summarizes the metabolic rift as follows: "Labor pounds and wheedles rocks and soil, plants and animals, extracting the molecular flows out of which our shared life is made and remade. But those molecular flows do not return from whence they came."[12] As crystallized in the metabolic rift, Marx's understanding of nature and its relation to production is an important starting point when theorizing reclamation and interrogating its terms and conditions. The epistemological basis of reclamation relies on a scientistic *disavowal* of the thesis behind metabolic rifts, promoting instead the perception that we as a species can carry on large-scale extractive processes without any serious, long-term damage to landscapes and ecosystems.

Embedded within reclamation's terms and conditions, including its gesture toward a vague concept of equivalent capabilities, is an anthropocentric—and indeed *capital-centric*—productivism that implicitly overrides the ecological complexities of pre-extraction landscapes. Jon Gordon elaborates on the implications of this anthropocentric, capital-centric productivism:

> This discourse of productivity asserts that the land will be more humanly useful, more profitable, because its productivity will be oriented to marketable ends . . . The amount of profit land can generate determines its value rather than diversity of life it supports, even if the latter must be sacrificed for the former.[13]

Reclamation, then, understands ecology primarily as a relation to capital, and in this sense it is consistent with its own logic when invoking "equivalent land capability." But such a narrow view of ecology—one that hinges entirely upon a landscape's profitability—reaches its limit when one begins to consider the material complexities of natural landscapes and ecosystems that reclamation, at least in spirit, hopes to mimic. This is precisely why it is essential to expand how we understand Marxist notions of labor, production, and use-values to include nonhuman animals. And while such a framing may read as a naive or superficial attempt to erode the boundaries between the human and the nonhuman, it is instead a fundamentally materialist gesture that begins to develop a politics through recognition of the role that nonhumans have played throughout history. Indeed, expanding notions of production to include nonhuman agents and actants is arguably a necessity when attempting to conceive of ecologically just relations in the twenty-first century. In other words, viewing nonhuman animals as producing for themselves expands and establishes grounds for a politics that moves beyond the anthropocentric confines of the Anthropocene.[14]

This anthropocentrism as it manifests in a privileging of the human, including the primacy of exchange-value over use-value in our current mode of economic organization, is a key factor in this epochal shift. Consistent with the types of posthumanist thought developed by theorists such as Donna Haraway, this move is not a betrayal of the spirit of Marx's understandings of production, nature, or labor.[15] In " 'Animals Are Part of the Working Class': A Challenge to Labor History," Jason Hribal complicates the anthropocentric assumptions behind conventional labor history, arguing that animals are, and historically have been, agents of production. "The basic fact," he writes, "is that horses, cows, or chickens have labored, and continue to labor, under the same capitalist system as humans."[16] Nicole Shukin pushes these observations regarding the role of animals in capitalism even further in her concept of "animal capital," as she traces the manner in which animals (including their labor and their commodification through processes of rendering) have been central and crucial to the rise of capitalism. If, following Hribal and Shukin, we begin to view animals both as *agents* and *victims* of capital that are central to its hegemony, as well as laboring beings in their own right, the concept of "equivalent land capabilities" becomes even more unstable, as the anthropocentric bases that form the foundations of the metrics of reclamation become ever more apparent.

Nature without Ecology: Ecomimesis and the Aesthetic Dimensions of Reclamation

If part of reclamation's problem lies in its material artificiality as a landscape stripped of its use-value and rebuilt in the image of an idealized form, another part of its problem lies in the aesthetic. Timothy Morton's work on environmental aesthetics helps in developing this point further. In *Ecology without Nature*, Morton establishes a conceptual apparatus (what he calls a "device") from which to critique dominant Romantic notions of nature as they work in the genre of nature writing and artistic production in general. He does so by developing the notion of "ecomimesis." Ecomimesis, Morton explains, is an environmental literary aesthetic that seeks to privilege, reflect, and embody nature—its ambience, its atmosphere—in its poetics, and in its weak form often reproduces the very same binaries it hopes to erode or erase.

But what does a largely literary aesthetic have to do with reclamation and its emphasis on innovation and progress? To view reclamation, and indeed the contemporary energyscape more broadly, as somehow outside the realm of aesthetics, is to misread its core imperative, especially in relation to the purpose of reclaiming natural, postextractive landscapes: to reconstruct an environment on the basis of a pernicious mixture of aesthetic and anthropocentric productivity. Although there are numerous potential end-uses that guide the reclamation process, there remains a residual emphasis on the aesthetic. "Equivalent land capability," in a self-sustaining and auto-productive manner, has yet to be properly achieved, if it is at all possible in the first place. This is precisely why Moore's evaluation of the success of reclamation centers primarily on "pretty *sights*," an anthropocentric way of experiencing nature that underscores reclamation's key drive *in the first instance* to aesthetically mimic that which has been internalized as "nature" in the cultural imaginary. In its privileging of the idyllic aesthetics of nature that can be traced to the types of Romantic conceptions of nature Morton elaborates upon and critiques, reclamation projects reinscribe the problematic dynamics of nature versus culture that privileges the latter over the former. While Morton's response is to call for an ecology without nature—an ecology that is not built upon the same problematic concept of nature that marginalizes it in the first place—reclamation projects invert this relationship in their mimicry of nature, producing a "Nature" *without* an ecology of any dynamism or vitality, fundamentally lacking in what Morton has elsewhere called "dark ecology"—the unseemly, "monstrous," and often brutal aspects of the natural world overlooked in Romantic idealizations of nature.[17]

The nexus of ecology, Romantic aesthetics, and scientism crystallized in reclamation exposes its status as an artificial landscape embedded in a colonial epistemological framework. It serves us better critically, then, to view landscapes like Gateway Hill as always already human-altered. Such viewing immediately destabilizes the Western idealized landscape and the politics of colonization in reclamation. Richard Grove's work, especially his *Green Imperialism* (1995) and *Ecology, Climate and Empire* (1997), is instructive in illustrating the colonial roots of conservation science. In both works, he traces the complex historical relationship among environmental conservation, environmentalism, ecology, and colonialism. My aim in underscoring this history here is not to dismiss the important work of the ecological sciences, but rather to historicize their Western epistemological origins in order to trouble assumptions that remain within ecology as science. Most important, we can read ecology dialectically by drawing distinctions between the ways in which ecology serves capital or, in the case of Grove's focus, the colonial apparatus en masse, instead of the peoples and wider environments that can be broadly understood as a kind of common. Such framing demonstrates the ways in which reclamation is materially and culturally—that is, aesthetically—wedded to the same dynamics that makes reclamation necessary in the first place.

Reclamation for What? The Symbiosis of Reclamation and Petrocapital(ism)

Who and what does reclamation serve? In her book *In Catastrophic Times* (2015), Isabelle Stengers labels those who serve Capital and, simultaneously, the destruction of the planet by deploying their knowledge—including financiers, scientists, politicians, and so on—as "our guardians."[18] There is a productive overlap between Stengers's guardians and Marx's notion of "general intellect," a term from the *Grundrisse* that describes the ways in which knowledge is deployed as a force to *reproduce* capital, operating as a kind of immaterial productive force. Through general intellect, Marx predicted the hegemonic role of knowledge as a productive force in the maintenance and reproduction of capitalism that serves the advanced stages of postindustrial capitalism, a stage of capitalism fueled by the types of energy intensification that fossil fuels arguably make possible.[19] In the case of reclamation projects, the knowledge deployed by "our guardians" as a kind of general intellect establishes the *perceived* possibility for a reconciliation of the contradictions of petrocapitalism. In other words, we can sustain our current fossil-fueled energy culture by leveraging science as general intellect to superficially eradicate the metabolic rift. We need no longer worry, the story goes, about the destruction of landscapes, about anthropogenic climate change, and so on, when our petrocultural guardians can mitigate these mere symptoms with "new ideas."

The role of science in oil sands reclamation and in the maintenance, expansion, and reproduction of petrocultures reveals the ways in which these dominant modes—science and (petro-)capitalism—often function symbiotically. Stengers's guardians name this pervasive and problematic relationship between benefactors of capitalism and its supporters (including some scientists) by collapsing seemingly heterogeneous factions of capital into a homogeneous group based on a single, shared, and constitutive effect: the self-justified furthering of the interests of capital, of extractivism, and of ecological destruction well into the twenty-first century and beyond.

Consider the land reclamation program, offered as a major leading to a BSc degree in environmental and conservation sciences through the University of Alberta's Faculty of Agricultural, Life, and Environmental Sciences. The University of Alberta, located in Edmonton, has a close historical relationship with the emergence of the oil sands as a viable site of (profitable) extraction, having pioneered many of the technologies currently in use in the oil sands. According to the University of Alberta, a potential career for graduates of the land reclamation major is "reclamation specialist."[20] ECO Canada, a professional resource whose aim and tagline is to "build the world's leading environmental workforce,"[21] describes the career of a reclamation specialist, featuring a profile of a "role model" who states that "a reclamation specialist's role is 'to create a win-win-win scenario, where industry, the environment, and landowners all win.'"[22] But, as I have argued above, such a scenario remains extremely uncertain, especially in the case of natural landscapes returned to "crown land" after successful reclamation. Such a process forms the dialectic of bitumen extraction: using up a landscape's resources, on the one hand, while superficially reconstructing them, on the other, providing the appearance of a reconciliation of the metabolic rift. It is an error, then, to view reclamation projects like Gateway Hill as a kind of postextractive procedure that absolves us of ecological responsibility; instead, it is more accurate to view such reclamation projects as akin to other infrastructures necessitated by energy deepening, such as Onkalo, a nuclear waste storage site featured in Michael Madsen's 2010 documentary *Into Eternity: A Film for the Future*. As a nuclear waste storage site, Onkalo is 5 km long and 500 m deep, and must remain undisturbed for a hundred thousand years. Oil sands reclamation and long-term nuclear waste storage, it follows, are two ends of a spectrum that ties energy to issues of representation. While reclamation hopes to represent the possibilities for the real-time absorption of the externalities of petrocapital, Madsen's film makes clear the full extent of the problem through the hyperbolic temporal scale necessitated by nuclear waste storage. Both face the same limits at vastly differing scales. Such projects are a prominent yet under-theorized aspect of our collective energy unconscious, as they bury the unseemly and dangerous byproducts of increasingly intensified energy deepening and of the demands of our collective energy cultures.

Conclusion

In reclamation, nature is simultaneously recast in terms of a material and aesthetic exchange value where "pretty sights" and potential forestry and logging capacities operate as metrics by which to judge its success. The *Oxford English Dictionary* has several definitions for reclamation, including "a reassertion of a relationship or connection with something; a re-evaluation of a term, concept, etc., in a more positive or suitable way."[23] Indeed, reclamation is a reassertion of a particular relationship between the human and nonhuman, one that asserts a techno-scientific mastery masqueraded as stewardship. Global anthropogenic climate change encapsulated in the concept of the Anthropocene is premised on a recognition of the ways in which *all* landscapes are now human-altered ones. Reclamation projects like Gateway Hill can thus be rendered as manufactured and artificial, as human-constructed and perhaps no different than an office building in any given metropolis. After all, both sustain and reproduce a form of capital that in turn sustains the social life that bitumen extraction enables.

I have previously tended to view reclamation projects such as Gateway Hill in terms of Freud's uncanny and Baudrillard's simulacra.[24] These frameworks, however, are limited—they describe the effect and phenomenological experience of reclamation rather than what reclamation *does* on an economic, ecological, and social level. Mobilizing the myth of our technological ability to rebuild damaged landscapes to "equivalent land capability," reclamation is a reinscription of settler-colonialist behaviors that see land as something to be managed, another form of primitive accumulation that Glen Coulthard argues has been fundamental to historical and ongoing processes of colonization.[25] As "new ideas" inoculate discourses of ecological destruction—geo-engineering is now "climate remediation," fossil fuel companies are now "energy" companies, pipelines are now "energy projects"[26]—questions about the possibilities of a postcarbon energy future are made visible at the level of language, discourse, and culture more broadly. Rather than climate or ecological "remediation," as I have argued, the techno-utopian processes embodied in reclamation are instead a kind of *capital* remediation—attempts to mitigate the ever-intensifying contradictions of petroculture and petrocapital, but always already falling short by misrecognizing the social and cultural dimensions of energy. Such an understanding of energy not only reveals the limits of purely technological solutions to the ever-intensifying climate crisis, but also lays bare the necessary foundations for building a postcarbon energy future based on just social and ecological relations. These just social and ecological relations certainly cannot be achieved without science, but not a science that is implicitly and explicitly subservient to capital. Instead, such a transition will be achieved with a science that is in service *to* species-beings and ecologies rather than markets, and *for* a commons rather than enclosures.

Notes

1. Canadian Association of Petroleum Producers (CAPP), "Dr. Patrick Moore on Alberta's Oilsands," Facebook, 9 November 2011, www.facebook.com/OilGas Canada/videos/574336350020/.
2. Moore's memoir, *Confessions of a Greenpeace Drop-Out*, traces his increasing disillusionment with the alleged radicalization of Greenpeace's mission in the 1970s, his eventual removal from the organization, and the process through which he became a "sensible" environmentalist. See Patrick Moore, *Confessions of a*

Greenpeace Dropout: The Making of a Sensible Environmentalist (Vancouver: Beatty Street Publishing, 2011).

3. See Kevin P. Timoney, *Impaired Wetlands in a Damaged Landscape: The Legacy of Bitumen Exploitation in Canada* (New York: Springer, 2015). Timoney develops a comprehensive critique of the viability of reclamation based on existing data, which he points out are extremely limited—so limited that conclusions in the affirmative or negative are hard to draw. He casts serious doubt on official narratives that confirm the viability of reclamation projects.
4. The semi-official narrative of how the oil sands came into being confirms this. During a research trip to Fort McMurray, Alberta, Canada, in 2014, a screening at the Oil Sands Discovery Centre of a short film on the history of the development of the oil sands emphasized the role of American entrepreneurialism in shaping bitumen extraction today. Larry Pratt's *The Tar Sands* confirms this narrative. Larry Pratt, *The Tar Sands: Syncrude and the Politics of Oil* (Edmonton: Hurtig Publishers, 1976).
5. Alberta Government, "Alberta's Oil Sands Reclamation," Alberta Government, 20 June 2016, https://web.archive.org/web/20160512075516/http://www.oilsands.alberta.ca/reclamation.html.
6. Alberta Government, "Alberta's Oil Sands Reclamation."
7. Notable works include Jon Gordon, *Unsustainable Oil: Facts, Counterfacts, and Fictions* (Edmonton: University of Alberta Press, 2015), which briefly comments on reclamation in a larger study on the oil sands; Timoney, *Impaired Wetlands* (2015); and *Fact or Fiction? Oil Sands Reclamation* (Drayton Valley, AB: Pembina Institute, 2008).
8. Alfred Schmidt, *The Concept of Nature in Marx*, translated by Ben Fowkes (London: NLB, 1973), 15.
9. Schmidt, *The Concept of Nature in Marx*, 27.
10. "Productivism" is a key term in the degrowth movement that identifies historical tenets of both capitalism's and socialism's respective projects of modernity; the notion of a "productivist imaginary" has been developed by Diego Andreucci and Terrence McDonough in their reading of degrowth theorist Serge Latouche to name the worldview that sustains productivism. See "Capitalism," in *Degrowth: A Vocabulary for a New Era*, edited by Giorgos Kallis, Frederico Demaria, and Giacomo D'Alisa (New York: Routledge, 2014), 62.
11. For an extended account of the metabolic rift and its relation to contemporary capitalism and climate change, see Brett Clark, John Bellamy Foster, and Richard York, *The Ecological Rift: Capitalism's War on the Earth* (New York: Monthly Review Press, 2010); and McKenzie Wark, *Molecular Red: Theory for the Anthropocene* (New York: Verso, 2015).
12. Wark, *Molecular Red*, xiv.
13. Gordon, *Unsustainable Oil*, xli.
14. The role of "anthro-" in the Anthropocene is a hotly debated one. See especially Jason W. Moore, *Capitalism in the Web of Life*, which argues against the notion of an Anthropocene and for the notion of a Capitalocene, properly accounting for the role of capital and *capitalists* in generating the epochal shift. Jason W. Moore, *Capitalism in the Web of Life: Ecology and the Accumulation of Capital* (New York: Verso, 2015), 169–92.

15. Haraway explains her reasoning in understanding nonhumans as productive beings: "The actors are not all 'us.' If the world exists for us as 'nature,' this designates a kind of relationship, an achievement among many actors, not all of them human, not all of them organic, not all of them technological. In its scientific embodiments as well as in other forms, nature is made, but not entirely by humans; it is a co-construction among humans and non-humans" (Donna Haraway, "The Promises of Monsters," in *The Haraway Reader* [New York: Routledge, 2004], 66).
16. Jason Hribal, " 'Animals Are Part of the Working Class': A Challenge to Labor History," *Labor History* 44, no. 4 (2003): 436.
17. Timothy Morton, *The Ecological Thought* (Cambridge, MA: Harvard University Press, 2010), 59–68.
18. Isabelle Stengers, *In Catastrophic Times: Resisting the Coming Barbarism*, translated by Andrew Goffey (London: Open Humanities Press, 2015), 29–34.
19. This historical process of energy intensification is termed by political economists of energy as "energy deepening." See Bernard C. Beaudreau, *Energy and the Rise and Fall of Political Economy* (Santa Barbara, CA: Praeger, 1999). In "Energyscapes, Architecture, and the Expanded Field of Postindustrial Philosophy," Jeff Diamanti explores the relationship between energy deepening and the shift to postindustrial modes of economic organization, claiming that energy deepening is "a crucial component of what [Rosalind] Krauss called the 'root cause' of postmodernism." See Jeff M. Diamanti, "Energyscapes, Architecture, and the Expanded Field of Post-industrial Philosophy," *Postmodern Culture* 26, no. 2 (2016).
20. "Bachelor of Science in Environmental and Conservation Sciences—Land Reclamation—Undergraduate Admissions—University of Alberta," University of Alberta, 27 December 2016, http://apps.admissions.ualberta.ca/programs/ah/ah040/lndre1.
21. "About ECO Canada," *ECO Canada*, December 27, 2016, http://www.eco.ca/about/.
22. "Reclamation Specialist: ECO Canada," ECO Canada, 27 December 2016, http://www.eco.ca/career-profiles/reclamation-specialist/.
23. "reclamation, n, " *OED Online*, http://www.oed.com/.
24. Jordan Kinder, "Sustainable Appropriation: Consumption, Advertising, and the (Anti-) Politics of Post-Environmentalism" (master's thesis, University of Northern British Columbia, 2013).
25. Glen Coulthard, *Red Skin, White Masks: Rejecting the Colonial Politics of Recognition* (Minneapolis: University of Minnesota Press, 2014), 12.
26. In the time that I have been working on this chapter, the URL of the government of Alberta's website for information on the oil sands has switched from "oilsands.alberta.ca" to "energy.alberta.ca."

PART II

Figuring Energy Culture

CHAPTER 7

Capitalism in the Corpse of a Whale

Ackroyd and Harvey

Stranded

Our first-ever expedition with Cape Farewell takes us across a wild stretch of ocean fondly called the "Devil's dance floor" where the current from the Gulf Stream meets the cold Arctic waters of the Barents Sea and swells reach 30 feet, and where all but the most hardened sailor turns a paler shade of green. Standing up is difficult and keeping anything down harder. The deck jumps. Floors slide. Thirty-six hours of plunging seas. Sleeping patterns go out the porthole. Listless bodies lie around the cabin. A rite of passage that we have no choice but to go through, taking us to the edge. The deliverance three days later is the High Arctic, a place itself on the edge of existence . . .

Quite how we got to be part of this roller-coaster adventure is a simple story. In 2003 we received a letter from David Buckland, founder of Cape Farewell, asking if we wished to be part of a crew of twenty artists, scientists, and educators sailing to Svalbard in the High Arctic to study the effects of global warming in this remote and icy polar region, and then to report, in our different ways, from the front line of climate change.

There is no easy response to this place. The extreme cold wraps tendrils of ice into our clothes, coats eyelashes with frost mascara, and delivers shocks of pain to exposed fingers and toes. The physicality of being here reduces the need for introspective thought. Walks across the tundra mean absorbing hours of slipping into another reality, scanning the surface of the ground like a detector searching for evidence of complete difference, engaging in detailed observations of moss formation, ice crystals, shattered stones, preserved bones, lichens, and ice boulders. Yet evidence of human intervention is plain to see. Blackened coal dust stains white ice in the Russian-owned, coal-mining town of Barentsburg; walrus skulls butchered for tusks scatter Moffen Island; shorelines are infested with fishing nets, plastic bags, and buoys; thousands of whale bones litter the icy beaches.

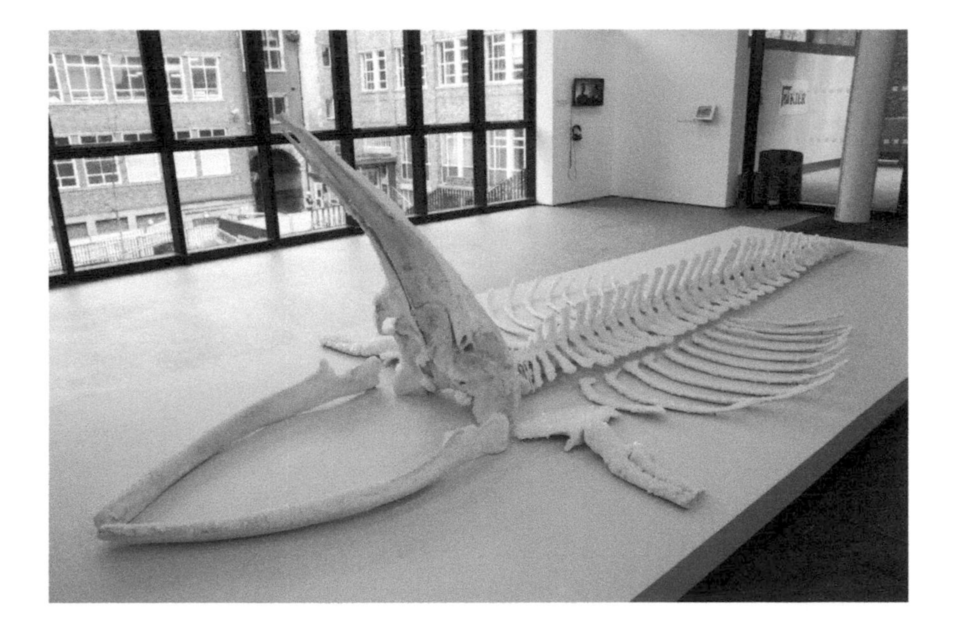

Back in the United Kingdom, an artwork emerges out of this extraordinary experience: a 20-foot-long whale skeleton encrusted with an ice-like growth of chemical crystals. Working closely with the Cetacean Stranding Programme at the Natural History Museum in London, we remove the skeleton from a minke whale washed up on the England's east coast. We clean the bones and then immerse them one by one in a highly saturated alum solution. It has the appearance of a fossil frozen in a crust of naturally brilliant crystals.

As the artwork progresses, so does our understanding of how in the last two hundred years the chemistry of the ocean has changed, altered for the first time in millions of years. Oceans absorb over one-quarter of all anthropogenic carbon dioxide entering the atmosphere, causing measurable declines in surface ocean pH, a process referred to as ocean acidification. This has a knock-on effect on organisms dependent on making calcium carbonate, damaging the corals, mollusks, and tiny zooplankton on which whales feed.

The exploitation in Svalbard brings home to us the scale of the slaughter of whales; in the nineteenth century there was worldwide demand for whale oil, now there is worldwide demand for petroleum. Both have taken a toll on the oceans. Both lie at the heart of capitalist enterprise and commerce. As we dissect the whale, picking at rotting flesh while at the same time admiring the work of maggots, the profound relationship between leviathan and industry takes hold on our imaginations, and we trace from shoreline to city how capitalism gestated in the corpse of a whale.

The must-have sporting accessory for the Edwardian gentleman striding onto the golf course was a club bag crafted from the 6-foot-long penile sheath of a blue whale. Clubhouse banter from one member to another apparently found much amusement from such an exclusive and chat-worthy prized purchase![1]

The maxim "let nothing go to waste" strikes a certain note when it comes to harvesting the material rewards of a whale. The leviathans of the sea, when hauled onto the whaling ship, would have had every last cetacean part used for one or other of the multitude of goods being produced across the industrializing nations of the nineteenth century: umbrellas, lipsticks, corsets, soaps, cosmetics, lubricants, insulin, glycerin, surgical stitches, tennis racket strings, paint, skin cream, drum skins, stock cubes, cattle fencing, iodine, liver oil, gelatin for the coating of photographic film, gelatin for jelly or for glue, fish bait, dog food, cat food, mahjong counters. The list stretches from shore to shore, port to port, where many a shore station near the whaling grounds became a city: Bergen, Boston, Brest, Buenos Aires, Darwin, Dundee, Durban, Glasgow, Hamburg, Hobart, Hull, Leningrad, Lisbon, Liverpool, London, Manila, New York, Philadelphia, Reykjavik, Rotterdam, San Francisco, Sydney, Tokyo, Vancouver, Wellington, Yokohama . . . "urban conglomerations plugged into the corpse of a whale, growing larger and larger by its light."[2]

The white translucency of light emitted from a candle made of "spermaceti" (the oil from the head cavity of the sperm whale) gave "the clearest and most beautiful flame of any substance that is known in nature."[3] In Britain, London became the best-lit city in the world, and "by the 1740s, five thousand street lamps were burning whale oil, expunging the primal darkness."[4]

Sperm oil was needed for heating and lighting all over the world. In the peak year of 1846 more than seventy thousand men were employed in the US whaling fleet alone, and the total value of whale products imported reached the "colossal figure" of $70 million.[5]

Capitalism and cetaceans are inextricably linked, a hostile takeover of leviathan proportions in which capitalistic intent scoured the oceans with scant regard so as to render four species of whale commercially, as well as almost biologically, extinct. The "right" whale earned its name from whalers who discovered that the thick blubber of the creature caused it to obligingly float on the ocean's surface after being speared to death; hence, it was duly decimated during the whaling heydays of the seventeenth, eighteenth, and nineteenth centuries.

The birthplace of capitalism can be traced to the Netherlands. Up to the Industrial Revolution, Amsterdam was the greatest trade city in Europe, home to a developing middle class of merchants and small manufacturers who thrived on trade, and home, too, to the first stock exchange and insurance companies. On the shores of this small nation the profit of capitalism emerged from the labor of the vast fleet of nearly six thousand ships that took to the high seas in relentless pursuit of whales. During the period from 1675 to 1721, the Dutch took 32,907 whales, which, at an average value of almost £900 each, brought a gross £30 million, a huge sum of money in that day and age.[6]

In the United States, during the period from 1835 to 1860, the annual imports of whale products averaged 117,500 barrels of sperm oil, 25,913 of common whale oil, and 2,323,512 pounds of baleen, valued in all at $8 million in that period.[7] America needed oil desperately at the time and was willing to pay handsomely for it. The great era of American whaling, when sailors took nearly 80 percent of world catch, ended before 1900. The collapse came not from the scarcity of whales, but from the discovery of oil in Pennsylvania in 1859, the year that Charles Darwin published *On the Origin of Species* and physicist John Tyndall gave conclusive proof of the radiant heat absorption of gases such as carbon dioxide, hydrogen, and methane, the so-called greenhouse gases that now plague Earth's atmosphere.

Notes

1. Sally Carrighar, *Blue Whale* (London: Gollancz, 1978).
2. Heathcote Williams, *Whale Nation* (London: Jonathan Cape, 1988).
3. John Adams to John Jay, 25 August 1785, in *The Works of John Adams, Vol. 8: Letters and State Papers, 1782–1799*, edited by Charles Francis Adams (Boston: Little, Brown, 1853), 309.
4. Philip Hoare, *Leviathan or, The Whale* (London: Fourth Estate, 2008).
5. T. Ivan Sanderson, *Follow the Whale* (London: Cassell, 1958).
6. Sanderson, *Follow the Whale.*
7. Sanderson, *Follow the Whale.*

CHAPTER 8

Tilting at Windfarms: Toward a Political Ecology of Energy Humanism and the Literary Aesthetic

David Thomas

Anglophone literary studies has thus far provided the emerging field of the energy humanities with one of its principal institutional seedbeds. To a disinterested observer, this particular conjunction of subfield and discipline might prove the cause of some surprise. It seems to forebode a distinct mismatch between concerns and methods. How much can literary scholarship really contribute to that larger interdisciplinary conversation that the energy humanities seek to stage, one defined by a focus on energy infrastructures, climate science, and policy-making? Frankly put, this skepticism is not entirely misplaced. Contemporary literature demonstrates precious little overt concern with the politics or material agencies of energy infrastructures. Imre Szeman crystallizes one of the foundational issues at stake by noting that those "who have begun to grapple with the cultural absence of oil in a period shaped by and around the substance" quickly discover that "few literary texts in the period of world energy literature address the power of energy to shape the world."[1] Reaching similar conclusions, novelist Amitav Ghosh finds that when he "tri[es] to think of writers whose imaginative work communicated a . . . specific sense of the accelerating changes in our environment," he is soon at a loss. Indeed, in Ghosh's estimation, one can count the exceptions on one hand.

Szeman and Ghosh are unusual in a field in which the more common tendency is to subtly overstate contemporary literature's attentiveness to the history and politics of energy and infrastructure. One can see this underlying tendency rise to the surface in many of the otherwise excellent essays that have sought to bring literary studies into contact with discussion of the

Anthropocene. In theory, formal analysis of the literary text always sheds light on the larger issue, but in practice it can often seem that we pause over close readings—lavishing minute attention on a narrative's relatively incidental references to infrastructural or energic issues—only to placate the disciplinary gods, or to prove that our signal professional practices have a distinct and significant role to play in the struggle to overcome the world-system's contemporary ecological travails. Although we like to position our close readings as the occasions when diligent attention allows us to unfold literature's virtuosic insights into infrastructural, ecological, or energic matters, candidly speaking, these are more often than not the moments when we smuggle the larger prospects into view through the back door. Yet "back door" is perhaps not the most apposite choice of metaphor, as it is more often the case that energy critics find themselves on the hunt, not for hidden and intentionally constructed thresholds, but rather for the incidental fissures in the surfaces of the text, for those moments when literary writing is rendered momentarily transparent to the vast infrastructural assemblages through which it has circulated in its efforts to arrest and beguile our attention. The glancing references to a turnpike or a gas station, the dinner party conversation that briefly touches on the subject of climate change: these become the kind of occasions on which the energy critic dwells in an effort to disclose the shadowy outline of the massive fossil fuel infrastructures that surround and sustain us as we read.

Yet our readings are often carried off with such skill and dexterity that all concerned can lose track of the heuristic struggles that generated them, struggles in which the critic is engaged in actively, if silently, resisting the prevailing conceptual currents that have carved out the thematic pathways of the text itself. Indeed, the motive force of the literary tradition generally seems intent on trammeling the reader's attention away from the massive social-machinic hyper-objects that energy humanism takes as its primary object of analysis, funneling the reader's attention back down toward the "rich" inner life of the cultivated individual. If we want to arrive at a better understanding of how to interpret literature under the black sun of climate crisis, then this tension—this sense that we find ourselves reading at the cross-currents of history, searching for new visions while prevailing currents carry us astray—is one that we must not leave undiagnosed.

This is a set of contentions that I shall return to in more depth as my argument unfolds—indeed, the overall interventional thrust of the chapter is already now in view—but I pause for a moment here to allow the signal concerns of energy humanism to become, in their own right, the object of a little analytical scrutiny. For at one and the same time as this chapter tries to account for contemporary literature's energy lacuna, it also attempts to grasp the institutional dynamics that have helped to promote the development of energy humanism as a distinct mode of inquiry. For it is my intuition that the movement is itself time-stamped by the particular academic environment and funding climate within which it has emerged, and it is my unfashionable suspicion that a self-reflexive account of this institutional environment will bring us a step closer to understanding the particular set of constraints and possibilities within which humanists now operate.

In some respects, these determinative conditions are all too familiar. In a contemporary context of global economic stagnation and welfare state contraction, we seem increasingly to be, as Sarah Brouillette and I have suggested elsewhere, "experiencing modernization of industry without modernity's attendant social forms: without, that is, the institutional, social, and cultural features associated with development, such as universal public education, democratic state institutions and, yes, somewhere down the line, a literary sensibility."[2] And in view of these broader trends, literary reading, like the opera or the art gallery, is now more and more clearly the preserve of expensively educated elites, many of whom attain this education through the assumption of an immiserating or indenturing debt load. It consequently means something different to read the literary novel today than it did in the age of the GI Bill, or the days of Jacques Rancière's journeymen scholars, or the era of cooperative working-class education movements, which is to say, in those days when it appeared that Raymond Williams's "long revolution" was still unfolding. And it is in relation to economic contraction, wealth polarization, and social state de-development that I think we have to acknowledge that most energy humanists are not brought to their work by entirely lofty and disinterested concerns: funding is a key issue here. In the broader context of a cash-strapped and revenue-hungry academy, funding bodies increasingly seem willing to loosen the purse strings only for humanistic research that trumpets its pragmatic applications.

This is not to imply, however, that we must view these new conditions as an unprecedented departure from the more enlightened norms that presided over the development of higher education in the days of the Keynesian dispensation. Indeed, and to the contrary, it is more likely that the changes we are observing partake more of the logic of ramification than of rupture. Literary pedagogy's instrumental function can be traced far back beyond the Keynesian campus to the inception of the literary field. Thanks to the many accounts we have now of the co-development of urbanization, industrialization, and mass literacy, it is relatively straightforward to show that the concept of the literary came into common currency precisely as elites attempted to manage, curate, and categorize the voluminous output of the industrial printing press. For example, in John Ruskin's polemics in favor of universal education, we see an early and fairly exemplary instantiation of a mode of managerial philanthropy that proposed that the bourgeois literary sociolect should be institutionally disseminated as a means of enfranchising, controlling, and "ennobling" the plebeian masses. Comparable visions have hardly been rare. Think of Thomas Babington Macaulay's notorious "Minute on Indian Education," or Matthew Arnold's "sweetness and light" as source-texts for faith in the civilizing, softening, humanizing impact of exposure to elite culture.

Yet, with all the necessary provisos in place, it remains evident that some aspects of our institutional experience are unique to our own moment. Indeed, it can hardly be a coincidence that the instrumental orientation of humanistic inquiry is becoming self-consciously explicit at precisely the same time as international governments' commitments to the promotion of the liberal arts is on the wane. The old faith seems to be in decline, and in this more skeptical policy-making climate, the celebration of literature's "uselessness"—long one of its special boasts—has become increasingly hard to sustain. In fact, it seems clear that the effort to buttress humanism's imperiled institutional position has seen cannier scholars and academic associations begin to index humanism's concern with "hard," pragmatic objects of analysis, in an effort to flag the enduring social utility of their work.

Pierre Bourdieu would define these responses to economic constraints and governmental initiatives as forms of "heteronymous" determination, and I mention them at this early juncture in an effort to follow through on Walter Benjamin's injunction to do "justice to the concrete historical situation of the interest taken in the object."[3] In appraising the "historical situation" of our interests, I emphasize that the rise of self-branded "digital," "medical," and "energy" humanities cannot really be candidly understood in isolation from the global economic maladies that have rendered states increasingly jealous custodians of the public purse, keen to ensure that public monies are apportioned to work that can make a more or less persuasive claim to social or economic "impact."

Whatever one thinks about these developments—and whether one prefers to explain them as reversible consequences of neoliberal ideology, or as the ongoing ramifications of a prolonged and deepening global economic malaise—there is an opportunity here that scholars concerned with climate crisis can ill afford to neglect. In ecological terms, the hour of need has long been upon us, and the newly utilitarian disposition of the humanist academy offers energy researchers a path to resources that they have little option but to take.

The situation has become so dire, and the need for collaborative knowledge exchange so belated and so pronounced, that we should probably avail ourselves of whatever platforms and conversational spaces global governance is willing to build and convene. As climate change becomes the stuff of our daily newsfeeds, it cannot but become an intensified area of concern for global governance's funding bodies, universities, and research hubs. And as an interdisciplinary focus on energy research is accordingly financially incentivized—often at implicit cost to people working in less utilitarian subfields—it becomes by default a kind of watering hole, to which many of us are drawn in our search for intellectual and professional survival.

The question of how best to make use of these spaces is one I shall return to, but for the time being I focus on trying to understand the kind of research paths that this funding environment has begun to open up. I take my own research trajectory as an anecdotal point of departure. There are doubtless pitfalls to this kind of approach—one is, for instance, always prone to overgeneralize the particularities of one's own experience—but I adopt this method, despite its risks, as a way of making the structural position of today's early-stage researchers explicit, and as a way of demonstrating how our particular institutional position affords us a keen insight into the developmental trajectory of the post-Keynesian university. Precisely because those on the academic periphery are more exposed to the new order's characteristic rigors and constraints, we are thrust—through the aegis of enlightened self-interest—into what Max Weber would define as a "prophetic" institutional role.

For instance, in the context in which I write—in a Canadian academic funding economy where money is increasingly allocated in relation to "Future Challenge Areas"—failure to respond to policy-making signals spells an increased risk of financial hardship. One is left with little option but to pick up the gauntlet. Yet the resultant effort to balance the discordant claims of Old World disciplinarity and New World instrumentality can lead the funding-hungry into some fairly contorted positions. One can, for instance, make a funding pitch around the claim that "contemporary literature has produced incisive accounts of the specific and concrete ways in which energy policies have shaped lives and organized nature," only to discover that, on closer inspection, this does not really prove to be the case. Yet given the broader institutional climate, the structural incentive to fudge one's findings is fairly pronounced. The entrepreneurial habitus of the contemporary funding economy encourages and elicits a general affect of can-do pragmatism, a milieu that proves much less receptive to the kind of naysayers and *enfants terribles* that thrived in the former culture of critique.

As my own research got under way, I quickly discovered that the paucity of literary works evidently interested in energy infrastructure forced me to the think about literary aesthetics in a relatively abstract manner. In trying to account for contemporary literary fiction's energy lacuna, I began to draw comparisons between literature and other cultural mediums, which seemed to furnish the researcher a more promising store of archival material. Indeed, contemporary literature seemed to provide the energy humanist a singularly awkward field of play.

Yet it is difficult to see precisely why this situation has developed as it has. After all, some of literature's most abidingly influential figures dabbled in the kind of cognitive mapping projects that energy humanists now prize. As Michael Rubenstein's recent rereading of *Ulysses* illustrates, James Joyce's writing consistently demonstrates a fascination with "the collective invention of the collective structures of invention,"[4] a vision that necessarily departs from the individually focalized viewpoints that still dominate literary fiction today.

In Rubenstein's reading, this element of Joyce's vision is never more clearly instantiated than in the novel's "Ithaca" chapter, a "moment in the text that has recently been rethought" by scholars such as Fredric Jameson and Joep Leersen "as a narrative crux."[5] This moment hinges on an apparently digressive reverie in which Joyce "takes the reader through a social infrastructure, the Dublin waterworks, that would ordinarily be considered 'non-narrative' . . . in any more standardized or traditional aesthetic practice."[6] In characteristically thorough and affronting fashion, Joyce moves through the system's technical and financial architecture with minute attention both to the precisely calculated infrastructural proportions of the project and to the methods of taxation by which the Irish state captured and redirected the capital flows that sustained it economically. Noting that the writing seems "plagued by passive voicing, tortured syntax, and proliferating punctuation,"[7] Rubenstein finds that "the trick here" is that this labyrinthine sentence "does not imitate the effortless flow of water; rather, it imitates the laborious flow of the waterworks: the whole complex of technology and labor for diverting water from its natural—'flowing'—course for utilitarian ends."[8] As Rubenstein describes this socioecological "complex," readers of Andreas Malm may recall his bravura account in *Fossil Capital* of the initial development of Britain's sprawling network of water-powered industrial cotton mills, a social-machinic assemblage designed around the water wheel's capacity to harness and transform the natural flow of water into the motive force—the fuel—of automated production. And, conversely, as one returns to Malm's work in the light of Rubenstein's reading of Joyce, one finds oneself keenly aware of how far Joyce went in anticipating the broader outlines of Malm's socioecological outlook. Indeed, as Malm describes how "the first act of the manufacturer, the mechanic or the millwright was to locate and appropriate a suitable force already existing in nature, the task of the prime mover being merely to harvest the force and pass it onto other bodies,"[9] the singular perspicacity of Joyce's literary vision stands out in bold relief.

Bourdieu's concept of the "collective invention of the collective structures of invention" serves Rubenstein well as he seeks to categorize the implicit politics of Joyce's "comedic" infrastructural aesthetics. But as Rubenstein turns to Bruce Robbins's concept of the "sweatshop sublime," we encounter an interesting cluster of categorical questions. Drawing on Robbins's influential formulation, Rubenstein argues that the overall result of the "Ithaca" chapter is to make visible the "social sublime" that trammels the trajectory of the novel's individual characters, a practice that foregrounds the collective structures of invention that sustain the life of Dublin's residents.[10] Yet given the kernel of "pragmatic utopianism"[11] that Rubenstein discovers in Joyce's representation of the waterworks—the sense that "in order to be effective in the political sphere [utopian thinking] *must* be grounded in realized social and political structures that must, in turn, be *re*invented"[12]—we might pause to consider whether Robbins's concept of the social sublime is really the one that we want to lean on at this particular juncture. For as Sianne Ngai reminds us in a recent interview, "for Robbins the sweatshop or capitalist sublime is an aesthetic response to the negative or meta-cognition of a cognitive and/or perceptual difficulty."[13] Ngai goes on to discuss the metanoic movement from insight to action that defines Robbins's original use of the term, wherein the overwhelming and appalling apprehension of planetary-scale systems of exploitation elicits a reorientation of the viewer's political outlook, one that can find satisfaction only in the pursuit of corrective political action.[14]

Yet in Joyce's representation of the infrastructures of postcolonial Dublin, we look onto a less dystopian prospect than Robbins's concept of the sweatshop sublime is designed to identify; indeed, we look onto an incipiently utopian set of socioecological relations. For as Rubenstein demonstrates, Joyce invites us to look onto Dublin's "public works" with an approving eye, offering us a "reimagining of the urban community that, while still shot through with the sense of loss and alienation that Jameson highlights, is nevertheless a powerful, hopeful, and original imagining of the urban community."[15] And by identifying the "Ithaca" chapter's latent ability to affirm the virtues of collective world-building—to disclose how the collective invention of the collective structures of invention can work to socialize as well privatize natural wealth—Rubenstein supplies us with a vital reading of an exhaustively studied text.

These considerations become crucial as we try to transpose Rubenstein's focus on public works infrastructure into the particular registers of energy humanism. Postcolonial Dublin's modernization program is much easier to figure in a utopian light than the sprawling and destructive assemblages of today's ecocidal and neocolonial fossil fuel apparatuses. Indeed, looking onto the latter prospects, one seems more prone to feel the crushing weight of "the totality of a system whose complexity and scale seems to fundamentally exceed the mental faculties we've been given to process it"[16] than to apprehend such vistas as sites where our capacity for collective agency is revealed.

Rather than conflate Robbins's notion of the "sweatshop sublime" with Bourdieu's understanding of the collective structures of invention, it seems that we might do better to position these two stances—the one that grasps the utopian potential of our collective socioecological self-making, and the other that asks us to defy the ecocidal and exploitative logic of existing infrastructures—as entangled in a dialectical relationship that structures and defines the politics of our geohistorical moment. For mired as energy humanists often are in disempowering contemplation of the petrocultural sublime—studying, for instance, the vile sublimities disclosed in Edward Burtynsky's aerial photographs of the Athabasca oil sands—the utopian and agential content of Joyce's infrastructure aesthetic is something we cannot afford to lose touch with.

Yet we should emphasize that the kind of affective and conceptual contributions that Joyce's work can make to this endeavor, are—with regard to subsequent literary trends—far from typical. In attempting to find prospects capable of either shocking us into political action, or reminding us of our capacity for collective socioecological agency, contemporary literature often seems more apt to serve as foe than friend. Only a handful of literary writers have developed Joyce's sort of holistic infrastructural vision with anything like his degree of intellectual rigor. This situation raises the question of why—given Joyce's ostensible importance to the field as a whole, and given the invitation that he extends to explore these strange manmade vistas in greater earnest—literary writers have opted to return again and again to what Rubenstein defines as "standardized or traditional aesthetic practice[s]."[17] And it also raises the question of what constitutes these "standard aesthetic practices" as recognizable literary forms.

If the adjudicatory statements of literary prize panels or the commentaries of high-profile literary critics are anything to go by, we could do worse than to begin our definition of "standard" literary practice in reference to the concept of "beautiful writing." For arbiters of literary value all too often justify their commendation of a particular text by explicitly celebrating the beauty or "elegance" of its composition. Perhaps this is the flip side to criticism's experience of Joyce's mimetic representation of Dublin's waterworks as a torturous, "tedious," and "laborious" departure from the literary mode. If this were the case, it might help to explain the puzzling characterization of infrastructure as "nonnarrative" material. An explanation does seem to be required, for in the ambit of the concept of the Anthropocene, the pressure of our moment's grand periodizing device directly invites us to consider the infrastructural development of fossil capital as the story of our time, as "narrative material" par excellence.

In trying to denaturalize contemporary literature's "standard practices," it can be helpful to contrast it with that other stronghold of bourgeois high aesthetics: contemporary art. Even to the unpracticed eye, it is immediately evident that the more barbed and affronting tendencies of the other field separate it from contemporary literature's marked preference for the beautiful. In critical appraisal of contemporary art, mentions of a work's beauty are decidedly scarce. While in the days of anglophone high modernism writers such as Ezra Pound, Gertrude Stein, and Percy Wyndham Lewis also experimented with a more angular and alienated affect, in the decades since, mainstream literary aesthetics has beaten a steady retreat back to a more soothing and neoclassical mode. Suffering subjectivity beautifully rendered has exerted a powerful gravitational pull across the entire field. To be sure, there have been challenges to the dominant taste formation—in the British context one thinks of the roiling postpunk aesthetics of Irvine Welsh or Nial Griffiths, or the knotty tumultuous panoramas of poets influenced by the Cambridge school, such as Keston Sutherland and Verity Spott—but it remains difficult to debate that there has been a marked divergence between the developmental trajectories and exploratory inclinations of the visual arts and literature. To brutally oversimplify: the former has tended to ugliness and exteriority, the latter to beauty and interiority.

While art installations and "nonsites" have increasingly engaged in elaborate conceptual mapping projects—in efforts to grasp and differentiate the sprawling assemblages of contemporary capitalism—the literary aesthetic has more often than not developed under the sway of the "interiorizing thrust" of post-Romantic literary expression. One could point to the systems novels of writers such as Thomas Pynchon, Roberto Bolaño, or David Foster Wallace, or the late modernist poetry of J. H. Prynne as exceptions. But, generally speaking, literary fiction has concerned itself with boring down "into" the interior life of the aspirational individual—"into" the conflicted interplay of libido, ethnicity, kinship, gender, profession, and nation that has defined the postwar summons to upward mobility. The intimate "I" of the professional subject, rather than the external "we" of the social-machinic hyperobject, has been established as the dominant focal point of literary fiction.

Another feature of the contemporary literary novel that constrains the energy humanist is that—compared to movies, television, and the visual arts—novels are much less inclined to pick up on incidental environmental information. Think, for instance, of the now ubiquitous drone flyovers that punctuate and frame the narrative action of television noirs such as *True Detective* or *Mr. Robot*—where the sprawling petroscapes of Louisiana and California, or the glowing nighttime technocanyons of New York City, return over and over as stunning visual refrains—and consider how few novels would attempt the same feat. While "the cut" has made the leap into literary fiction, "the flyover" generally has not. And could we stomach it if it had? Contemporary critical reaction to Joyce's "waterworks" suggests not. This double bind renders today's literary fiction a somewhat inhospitable environment for the energy humanist: on the one hand, there is much less "room" for accidental and incidental infrastructural information in print narrative; and on the other, there is little appetite for a deliberate focus on the same. Even older experiments in such a vein are still received with puzzlement and aversion.

My point is not that literary writers have inexplicably declined to tell the signal story of our moment, but rather that the curatorial norms, and the prevailing doxas and taboos, of the contemporary literary prize economy appear to have rendered the relevant source material somewhat verboten.

The kind of concerns I have in mind here surface in the commentary that has surrounded environmental scientist Vaclav Smil's now somewhat notorious statement that "timeless artistic expressions show no correlation with levels or kinds of energy consumption: the bison in the famous cave paintings of Altamira are not less elegant than Picasso's bulls drawn nearly 15,000 years later."[18] Responding in a recent paper, Szeman acknowledges that "Smil is right to caution against crude 'energy determinism' in relation to historical development," before going on to counter the overall thrust of Smil's argument:

> the claim that energy has no impact on literature and culture is unsustainable, whether we understand this impact in a narrowly material sense—in the very substance of the acrylic paints used on modern canvases, the stock used to shoot films, or in the electricity required to run

> printing presses and generate electronic signals—or socially, through its figuration of social capacities and expectations.[19]

On an empirical level, it is hard to debate the merits of Szeman's rejoinder. And it would be easy to assume that this particular disagreement is a product of the kinds of misunderstandings that are prone to arise at the interstices of academia's two cultures. Yet what intrigues me about Smil's argument is the extent to which it simply reflects back to cultural criticism the kind of reading practices and aesthetic doxas that its most prestigious canonizing intuitions continue to propagate.

Indeed, Smil's remark—intentionally or not—harkens back to T. S. Eliot's similar claims before the cave paintings in Périgueux, which prompted him to conclude that "art never improves."[20] Unlike Smil, however, Eliot went onto to modify this claim, and in essays such as "Tradition and the Individual Talent" he balanced the aestheticist belief in art's transhistorical qualities with the recognition that its signature forms and modalities were also continually subject to historical transformation and remediation. Eliot's reading of the historical development of art—his assertion that the concept of "art" partook of an intrinsic historicity—relied on a broadly conservative belief in the great tradition, one that insisted that the artist must be a scholar of the canon, so immersed in the nuances of high culture's historical development that his or her contributions to the field would always implicitly recapitulate and respond to the vast corpus of exemplary works that had gone before it: "The existing monuments form an ideal order among themselves, which is modified by the introduction of the new (the really new) work of art among them."[21] In stressing the value of "the really new," Eliot, in his conception of aesthetic excellence, gestured toward what he dubbed the "simultaneous order."[22] In this schema, the addition of "really new" work changed one's understanding of the relationships between all the earlier works as well: "the past should be altered by the present as much as the present is directed by the past."[23] With this kind of conception of the historical development of culture, we have apparently taken a step beyond the notion that the "timelessness" of art renders it immune to the vagaries of history.

Yet, when subjected to scrutiny, the distinction between "timelessness" and a "simultaneous order" can become a little hard to grasp. If art and literature are primarily defined by the play of great works refracted and reflected in each other, then attention to the shifting play of internal relations that prevails within this "ideal order" remains the primary mandate of cultural criticism. Intriguingly, there are moments when Bourdieu's and Eliot's understandings of the aesthetic seem improbably compatible. Mirroring Eliot's conception of art's "simultaneous order," Bourdieu's similarly finds that "when a new literary or artistic group imposes itself on the field, the whole space of positions and the space of corresponding possibilities, hence the whole problematic, find themselves transformed."[24] But what kind of light can such intensely field-specific modes of reading and interpretation hope to shed on the urgent ecological and political concerns of our own moment? If art and literature talk primarily of themselves and to themselves—and if this conversation is generally conducted in the well-furnished quarters of a relatively small bourgeois elite—is Smil wrong to imagine that the "aesthetic" falls, to all intents and purposes, "outside of history"?

That said, we would probably do well not to rush to judgment on this matter. Indeed, we might pause for a moment to consider that the near identical stance that Smil and Eliot take up before these prehistoric cave paintings might, in itself, be a phenomenon worthy of a little scrutiny. It seems quite conceivable that their experiences are less a result of the intrinsically agential properties of the images, and more a product of the educational and social milieu in which both men were raised, a habitus that took time—amid the busyness of more utilitarian concerns—to tutor them in the ameliorative virtues of the "pure gaze." Indeed, in their comments on the cave paintings Eliot and Smil both imply that their encounter unfolded according to that familiar set of subjective protocols that constitute the "pure gaze" as a distinct mode of perception:

> The aesthetic attitude is characterized by the concentration of attention (it "*frames apart*" the perceived object from its environment), by the suspension of discursive and analytical activities (it ignores the sociological and historical context), by disinterestedness and detachment (it sets aside past and future preoccupations) and, finally, by indifference to the existence of the object.[25]

As this mode of perception is translated from the museums where it was first developed into the "wild," we see the "pure gaze" effecting a "rupture which, by tearing the works out of their original context, strips them of the diverse religious or political functions and thus reduces them, by a sort of active *epoché*, to their properly artistic function."[26] And thus, as Szeman raises questions concerning Smil's judgment of the Altamira paintings, he is in an implicit and indirect sense struggling with the very heuristic strictures that the pure gaze imposes—"framing in" rather than "framing out," stressing sociological and historical context rather than suspending the discursive faculties, and introducing past and future preoccupations in place of disinterestedness and detachment.

That administrative pressure on the contemporary humanist academy now seems to be encouraging literary humanists to resist the very doctrines that their own disciplines have spent the better part of their existence propagating is a fascinating development, one that would seem to see the signature contradictions of our moment penetrating into the inner life of scholastic high-cultural experience. There seems to be less time, and less space, for the disinterested delectation of art's "simultaneous order." One might be tempted to respond to these observations with another round of lamentation about neoliberal philistinism. But if, as Bourdieu argues, the aesthetic has functioned less as an agent of dissent and emancipatory class antagonism, and more as a mode of bourgeois class consolidation and symbolic domination, then perhaps its retreat heralds the arrival of a more fragmented and protean set of cultural conditions.

I have already touched on the idea that there may be ways in which ostensibly lamentable policy-making developments—destructive of much of what initially drew us to our work in the first place—are also opening up new zones of encounter and intellectual exchange. I now want to suggest that we should be careful not to too readily conflate our desire to protect literary humanism from the shears of the de-developing social state with the effort to overcome the ecocidal energy appetites of the global economy. It is by no means self-evident that the old aesthetic doctrines should count on surviving the "transition" to a post–fossil fuel energy system. If this unlikely day of deliverance ever does materialize, we may be surprised to find ourselves watching—with a little of the disinterest of the pure gaze—as the sun sets on the old disciplines, taking with it the high-aesthetic, high-literary decadence of the Old World, and leaving behind a stranger set of aesthetic coordinates that have more to do with the "interested" functionalism of design.

The concepts that we need to imagine our current predicament come not from the self-referential "determinate reveries"[27] of the aesthetic, but from climate science, emerging from capital's social-machinic computational labs, as the discourses and technologies of systems theory and cybernetics have redefined policy-makers' understanding of the ecological agency of the global economy. From the vantage point of the world's climate science laboratories, it was becoming evident that the account of nature given by the natural sciences was not so much an account of an external order, as of an evolving system observing itself. As the scientific consensus on global warming crystallized, data on the climatic impact of carbon emissions were presented as a form of negative feedback, one that policy-makers were urged to use in a concerted effort to correct the developmental trajectory of the world-system. Though such projects are often construed in terms of "saving" nature from capital, the goal here is actually the techno-scientific regulation of the biosphere, a goal that would be better described, in Marx's terms, as the real subsumption of the world ecology.

Yet without the benefit of the economic dynamism that the world economy enjoyed in the years of Camelot, the prospect of undertaking a massive infrastructural transition away from carbon dependency seems stomach-churningly distant. This fundamental situation points us to some of the key conflicts and concerns that currently structure the global elite's internecine struggles, as the attempt to cybernetically regulate the biosphere—through the adoption of a more carbon-neutral energy base—is hindered at every turn by the inertia of an economic system that cannot as yet part itself from deep capital investments in the fossil fuel industry without, at the same time, further imperiling the dynamism of a stagnating global economy.

Can cultural consciousness-raising in elite circles help us to get beyond this impasse? Or will an intensified high-cultural concern with the Anthropocene and the social sublime simply leave us in the position of mapping the mappers, aestheticizing the outlines of an unfolding catastrophe that the administrators of the world-system—despite their efforts to the contrary—seem unable to forestall? Whatever the case, we can be certain that the developmental trajectory of curatorial and critical practice will prove a good deal easier to correct than that of our energy infrastructures. And thus, while cultural exploration and representation of energy assemblages and ecological systems seem likely to proliferate in the coming decades, we should not assume that this intensification of high-cultural concern will have any necessarily decisive impact on the infrastructural composition of the global economy itself. Perhaps we would take a step nearer to the truth of our predicament if we began by admitting that—absent more radical and concerted forms of political agency—all the biennials and essay collections in the world will do little to keep ecocide and genocide at bay.

Surveying the overall picture, we have much reason to be pessimistic—indeed, as Malm writes, "the prospects are dismal."[28] And this situation risks insinuating a subtle cynicism at the heart of any literary research project that seeks to orient itself in relation to energic issues. For my own part, I have found that the risk of exercising bad faith is pronounced, and I have often experienced an awkward sense of ambivalence about the work itself. Some years into the research, I remain a little confused as to whether I act—as I would like to imagine—on the side of the angels. Is it not more likely that I simply adopt such concerns as a means to an end, as a route to accumulating sufficient cultural capital that I will be shielded from the harsher effects of the ramifying ecological and political carnage? That said, one does experience occasions when the work itself, or encounters with networks of committed energy researchers and battle-hardened environmental activists, extends momentary but compelling invitations to that old devil hope. At such junctures one feels, as Malm again puts it, a pull toward Benjamin's "conception of history," toward a notion of political struggle that "draws its inspiration from the heritage of the oppressed in order to derail the ultimate disaster of the present."[29]

But before concluding that the overall effect of my line of argument has been to render contemporary literary fiction "irrecuperable" for energy humanism, we might look to the radical departure from prevailing critical norms articulated by Pierre Macherey. Conceiving literature as a "determinate reverie," he argued that it

> reveal[ed] and actively contribute[d] to certain fracture lines which run deep into historical reality and into the forms in which that reality is lived, imagined and represented. Seen from this angle, literature is no longer a matter of pure aesthetic creation, but becomes a form of knowledge, the material bearer of certain truth effects which require deciphering; and it is these truth effects which justify the interest we bring to it.[30]

Seen from this angle, the writer's point of departure is historically determined, and he or she writes drawing on the discourses and fictional complexes that animate the social imaginary at the moment of the work's composition. In explaining the twists and turns of the text's "coiled line," Macherey's critic is always in the grip of centrifugal forces, finding himself thrown back out toward the ideological currents and historical circumstances that had first supplied the text with its conditions of emergence, "probing [the text] . . . for those neuralgic points at which it betrays the shadowy presence within it of conflicting historical powers."[31]

In the ambit of energic and infrastructural concerns, this approach to close reading seems especially vital. As recent work by energy historians such as Malm and Timothy Mitchell has so clearly demonstrated, the development of capitalism's energy infrastructure was in large part determined as workers and industrialists vied for control of the energy flows that powered industrialization, conducting a series of slow-burning technical and logistical struggles—a dialectic of sabotage and automation—that played out over the course of centuries. We can discover the "shadowy traces" of these clandestine conflicts in literary writing, but to do so we generally already have to have some idea of what we are looking for. Without reference to other modes of research, exhaustive study of the canon's hall of mirrors brings little of this struggle immediately into view; rather than a lamp that sheds light on the nature of the world-system's subterranean ecological and political unconscious, literary works are more often than not a site where the outline of this agonistic world appears in shadowy relief, in the silence of what the texts appear not to say.

This is an assertion so obvious that it borders on redundancy, yet I include it here—even at the risk of appearing a little flatfooted—in an effort to underscore the peculiar awkwardness of the heuristic projects that geohistorical circumstances are now compelling the state's cultural custodians to undertake. The effort here—and I concede it may not be a welcome one—is to make this very awkwardness the subject of discussion in its own right. In doing so, I aim to draw attention to the kinds of heuristic fault lines that are being thrown up as the inertia of the old disciplines collides with the ferociously intense instrumentality of the de-developing social state. These fault lines have produced the academic terrain that emerging scholars are forced to navigate as they attempt to hone a Janus-faced mode of inquiry that is both responsive to the overarching interdisciplinarity of the prevailing funding structures, and "disciplinary" enough that it can negotiate the concerns of search teams and PhD committees that are—for the time being—afforded the structural luxury of positioning themselves as embattled defenders of the old verities.

Thus, a self-reflexive account of contemporary academic trends can seem to do more to disclose the structural nature of our professional and political situation than a retrospective study of the contemporary literary canon itself. Indeed, the study of literary writing in the age of the Great Acceleration generally seems to offer more insight into the dreams and fantasies that permeated elite culture as the signature modalities of petrocapitalism developed than into the underlying nature of the project itself. And if literature's determinate reveries are, as Macherey argues, a guide to the evolving nature of the bourgeois social imaginary, then it seems evident that, as the pipelines, refineries, and petroleum-powered assembly lines of petrocapital spanned out across the planet, elites took little time to ponder the changing nature of human "species being." Indeed, in their reflective moments they were more likely to be found imagining themselves as participants in a noble "simultaneous order," one that called back and forth to itself across the centuries. All the while, the "collective structures of invention" that constituted everyday life were exerting forms of ecological and climatic agency that the architects of these determinate reveries rarely paused to figure.

Yet how do we break with the strictures of this social imaginary? How do we rediscover—in this most dystopian of academic and geopolitical climates—a capacity for the kind of utopian vision that Rubenstein's reading of *Ulysses* brought to the fore? One could argue that a key—if not utopian—break is already under way, one initiated not by the agents of an aesthetic avant-garde but by international government itself. Despite its pitfalls, this may not be the worst line of argument. Although prevailing academic trends are already proving unevenly hostile to critical thinking, the results of the new funding frameworks may be set to become more fascinating, and their potential more manifold and unpredictable, as humanists, social scientists, and STEM researchers continue to converge—in increasing numbers—on the topic of energy.

The concept of energy itself sometimes seems to crackle with vast, untapped utopian potential. Most participants in the field, however beaten down by the unremitting drip-feed of dire datasets and dispiriting news cycles, will likely have experienced moments when a similar sense of things overtook them. So while there is good reason to fear the quietly coercive nature of the power dynamics that have presided over this process of convergence, in the grip of the process itself, it can still sometimes seem that the old Socratic injunction has at last begun to take on socioecological dimensions. The pressure of climate change—the negative image of capital's thirst for energy—is compelling us to "know ourselves" as creatures made out of infrastructures, energy and capital cycles, and ancient (and now disturbed) biorhythms.

Though rare, these moments of hesitant hope are valuable. Although the energy humanities are but one example of humanism's much broader turn toward overtly utilitarian concerns, their focus on energy still marks them out as an especially significant case. It is hard to conceive of another concept that can lay claim to such intrinsic profundity, to such ancient vintage, and to such pressing contemporary concern. In the course of a single conference, we routinely look on as the meanings and connotations of energy begin to fork off in every conceivable direction. In tracking its diverse implications and applications, we find ourselves tarrying in every place from bitumen sands to the sites of ancient animistic rites; we are returned to Athens and the formulations of first philosophy, and arrive in Geneva at the panels of particle colliders and nuclear reactors. The avenues branch out this way and that, yet whether we circle back via the secret lives of plants, the second law of thermodynamics, or the scripts of cuneiform tablets, the question (or is it the summons?) of energy remains a point of departure that we can return to again and again, knowing it as if for the first time. Faced with such dynamics, it is tempting to think that the concept of energy itself is uniquely capable of serving as a lodestar for the kind of synthetic vision we would need if we were ever to produce forms of social life capable of submitting the power of techno-science to a kinder and more holistic understanding of our place in the world.

So, although we as yet find ourselves unable to break through the energy impasse,[32] we should still pay careful attention to the profound changes taking place within late capital's knowledge production networks, changes that are incrementally shifting the content of our debates and the nature of our audiences. We begin to talk as much with engineers, scientists, and technicians as with other humanists and social scientists. Surely, there are opportunities here. Insofar as questions of energy and climate change are adopted as focal points of interdisciplinary knowledge exchange, these new forums present themselves—disingenuously or otherwise—as the staging ground for a radical reconstruction of intellectual and infrastructural life. Granted, given the lateness of the hour and the surging popularity of reactionary politics, it stretches credulity to the limit to place undue hope in these developments. But perhaps it has always been the case that only by pushing beyond the limits of caution and good sense can we ever succeed in bringing utopian prospects into view in the first place. A sure sign that such prospects are in view again will be that our arguments work less to justify the perpetuation of the old orders and interests—literary or otherwise—and more to understand and respond to the dilemmas that define our geohistorical situation.

Admittedly, the viability of this particular form of hope takes a wager on the ongoing ramification of late capital's systemic frailty, assuming that if present trends continue unabated—assuming, that is, that capital remains locked in the downward spiral of its own self-unmaking—the battle in the coming decades will be less to "overthrow" the old order than to survive the chaotic fallout of its auto-abolition. Such assertions doubtless overstep the limits of caution and good sense, but they at least have the merit of helping to undo the paralysis of political imagination that has characterized these recent decades, when it has so often been said that it was easier to imagine the end of the world than the end of capitalism. It is evidently no longer quite so difficult to imagine the latter. The vultures are already circling. And if we are to succeed in fending off the rise of the fascistic and absolutist pretenders that already wait in the wings, we will need to develop a more charismatic set of political and infrastructural alternatives than are currently on the table. Perhaps it is here that the better instincts of the new funding frameworks will find their lasting validation, insofar as their conceptual legacies survive the wreck of fossil capital, transmitting viable infrastructural exit strategies to those who survive the flood. In the meantime, we may have little option but to go on tilting at windfarms, acting in the quixotic conviction that other, wiser worlds will eventually begin to drag themselves from the husk of this dying one.

Notes

1. Imre Szeman, "Conjectures on World Energy Literature: Or, What Is Petroculture?" *Journal of Postcolonial Writing* 53, no. 2 (2017): 277–88.
2. Sarah Brouillette and David Thomas, "Forum: Combined and Uneven Development," *Comparative Literature Studies* 53, no. 3 (2016): 511.
3. Walter Benjamin, *The Arcades Project* (Cambridge, MA: Harvard University Press, 1999), 391.
4. Pierre Bourdieu, "Pour un savoir engagé," *Le monde diplomatique*, February 2002, 3.
5. Michael Rubenstein, *Public Works: Infrastructure, Irish Modernism, and the Postcolonial* (Notre Dame, IN: University of Notre Dame Press, 2010), 49.
6. Rubenstein, *Public Works*, 51.
7. Rubenstein, *Public Works*, 53.
8. Rubenstein, *Public Works*, 53.
9. Andreas Malm, *Fossil Capital: The Rise of Steam Power and the Roots of Global Warming* (London: Verso, 2016), 37–38.
10. Rubenstein, *Public Works*, 58.
11. Rubenstein, *Public Works*, 48.
12. Rubenstein, *Public Works*, 48.
13. Sianne Ngai, "Critique's Persistence: An Interview with Sianne Ngai," *Politics/Letters*, 27 February 2017, http://politicsslashletters.org/2017/02/critiques-persistence/.
14. Ngai, "Critique's Persistence."
15. Rubenstein, *Public Works*, 92.
16. Ngai, "Critique's Persistence."
17. Malm, *Fossil Capital*, 51.
18. Vaclav Smil, "World History and Energy," in *Encyclopedia of Energy Vol. 2*, ed. J. Vutler Cleveland (Amsterdam: Elsevier, 2004), 559.
19. Szeman, "Conjectures."
20. T. S. Eliot, "Tradition and the Individual Talent," in *The Sacred Wood and Major Early Essays* (Mineola, NY: Dover Publications, 1998), 29.
21. Eliot, *Sacred Wood*, 28.
22. Eliot, *Sacred Wood*, 28.
23. Eliot, *Sacred Wood*, 28.
24. Pierre Bourdieu, *The Rules of Art* (Stanford, CA: Stanford University Press, 1996), 234.
25. Bourdieu, *Rules of Art*, 286.
26. Bourdieu, *Rules of Art*, 294.
27. Pierre Macherey, *A Theory of Literary Production* (New York: Routledge, 2006), 363–64.

28. Malm, *Fossil Capital*, 394.
29. Malm, *Fossil Capital*, 394.
30. Macherey, *A Theory of Literary Production*, 363–64.
31. Terry Eagleton, "Preface," in Pierre Macherey, *A Theory of Literary Production* (New York: Routledge, 2006), x.
32. Dominic Boyer and Imre Szeman. "Breaking the Impasse: The Rise of Energy Humanities," *University Affairs*, March 2014, https://anthropology.rice.edu/sites/g/files/bxs1041/f/The%20Rise%20of%20Energy%20Humanities.pdf.

CHAPTER 9

Embodied Actants, Fossil Narratives

Maria Michails

Interviewed by Andrea Zeffiro

Maria Michails is known internationally for her interactive installations that link resource extraction and consumer demand with environmental problems specific to place. Michails's installations most often center on particular objects chosen in part for their nostalgic reference, but also for their citation of industrial history. The artist appoints these objects, such as a boat, a train, or an automobile, as entry points through which participants are invited to physically interact within the installation. Thus, participation by the audience is pivotal because the physical energy input by a participant is what generates the electricity required to activate an installation. Along these lines one might describe Michails's body of work as *co-productive*: direct and physical engagement from the audience is compulsory in order for the work to be energized, so to speak. By placing participants in the work and by asking them to activate specific installation pieces, Michails demands from participants a "thinking with" these objects in the context of a larger installation, and also in relation to energy regimes. Powering a motor, for instance, isn't an action aimed simply at generating electricity required for an installation. Rather, the objective is to insert the participant into the process of energy generation and consumption, and to ask them to consider their part in the process itself. Michails's work underscores energy as bound to nature and resource extraction, but it also asks that we reconsider energy along philosophical and ethical lines.

What follows is excerpted from a conversation that took place over four months in the fall of 2016, shortly after Michails had attended the Banff Research in Culture (BRiC) residency. It was there that Michails initiated a new work, *Mapping the Terrain: Pipelines, Bomb Trains, and the Narrative of Land*. This work examines energy infrastructures more closely by emphasizing the interwoven narratives between human and nonhuman actors affected by cross-border oil transport. Interestingly, as we discovered through our conversation, Michails's current preoccupation with energy infrastructures is a thread woven throughout her body of work, which has consistently drawn attention to the inadequacies of our energy regimes and their toll on communities, ecological processes, and nonhuman life forms. Rather than putting forward a succinct trajectory of an artist's output, our conversation theoretically orients Michails's practice so that we may better understand the capacity for artistic work and research to address complex issues pertaining to petroculture and the ways in which we come to understand our personal connections and orientations to it.

Andrea Zeffiro (AZ): I'd like to start by reflecting on your art practice and its commitment to a critical engagement with petroculture. I see thematic consistencies with respect to energy, but also more broadly in the referencing of resources in crisis. Can you explain the thematic connections across your projects and reflect on how you've come to a focus on petroculture?

Maria Michails (MM): For the last decade, I have been building installations that are powered by participants operating a mechanical, kinetic sculpture in the form of a boat, a railway handcar, or a simulation of an automobile. These human-powered sculptures generate electricity to activate installations composed of backlit photos, small video monitors, pumps, electronics, or other 12 volt devices. The impetus to involve the gallery visitor—who becomes a "performer" in the work—was to encourage an embodied understanding of energy production, fossil fuels, and resources in crisis, such as water or topsoil, specific to place. For example, energy and water in the desert in the project *EMERGY*, or energy and topsoil depletion in the Prairies or Midwest with the project *S.OIL*. Often we don't recognize how interconnected and interdependent these systems are. Therefore, we become removed from not just the processes but also the waste generated, because these processes are not readily visible. What if we had to generate our own electricity, through our own labor, in order to have water on demand, or even hot water? Why does it take fossil fuels to grow our food? What else is involved to bring that food to our table? How much energy is required? Are there alternative approaches?

The first project to utilize these themes was *EMERGY* (2008), where the boat was used as a motif to represent early historical mobility and trade. In *EMERGY*, community participants rowed the boat for twenty minutes each to power backlit photos of the Salt River that meanders through Phoenix, Arizona, and the adjacent cities. In 2010, I re-created the images to represent

Fig. 9.1. *EMERGY*, 2008. Exhibited at Harry Wood Gallery, Arizona State University, Tempe.

the city of Las Vegas and Lake Mead, where the Hoover Dam provides a significant portion of the electricity for Las Vegas. The water level of the lake has been receding in recent years, and it is threatening the dam's ability to produce power, shifting the demand to more coal-fired power.

In the second project of the human-powered series, *S.OIL* (2012), I focused on agriculture's dependence on crude oil and the paradox of using petroleum-based products to grow corn for ethanol. For this project, I built a railway handcar to represent the history of the train as an important form of transportation for people and goods. The train was pivotal to the expansion of cities during the Industrial Revolution and for bringing farm labor out of the city to farms (mostly in the United States). The rail system enabled the movement of manufactured goods to ports but especially encouraged large-scale, high-output farming and, in recent years, crude oil and ethanol. Ironically, this last oil boom (2009–14) and the increased circulation of crude by rail have created a backlog for farmers trying to get grain to market. Farmers now compete with the oil industry for train transport. Oil is extracted from their land, which they lease to the oil companies. I'll talk more about this a little later with the new work begun during the BRiC residency.

Fig. 9.2. *S.OIL*, 2012. Exhibited at Art Gallery of Southwestern Manitoba, Brandon.

S.OIL made visible the paradox of using fossil fuels to grow corn for ethanol, a very inefficient use of energy to make energy. The project highlights topsoil erosion and how detrimental it has been because of large-scale corn production, but it also presents an alternative type of agriculture, a potentially sustainable method of agriculture not requiring fossil fuel inputs. The plants in this project are hybrid grains (wheat, sorghum, and sunflower) developed by the Land Institute. Unlike the annual grain seeds (currently used in our agricultural systems), the hybrid perennials grow deep roots, enhancing and promoting topsoil regeneration, and are more resilient to climate change effects. Without the gallery visitors coming to pump the handcar, the plants would not get watered. The onus is on the community to keep the plants alive, a gesture that brought many issues to the foreground in the rural community where it was exhibited.

The Petri Series: Benzene (2013) project also concerned oil and, specifically, the hydrocarbon benzene, which is a known carcinogen (the microscopic cell images in the petri dishes are different types of cancers associated with benzene). Benzene is the sweet-smelling liquid in gasoline and a diluting agent in bitumen (oil sands) to make it flow better. But first let me say a few words about energy. Energy in the form of electricity is generated via

Fig. 9.3. *The Petri Series: Benzene*, 2013. Exhibited at Art Gallery, Central Michigan University, Mount Pleasant.

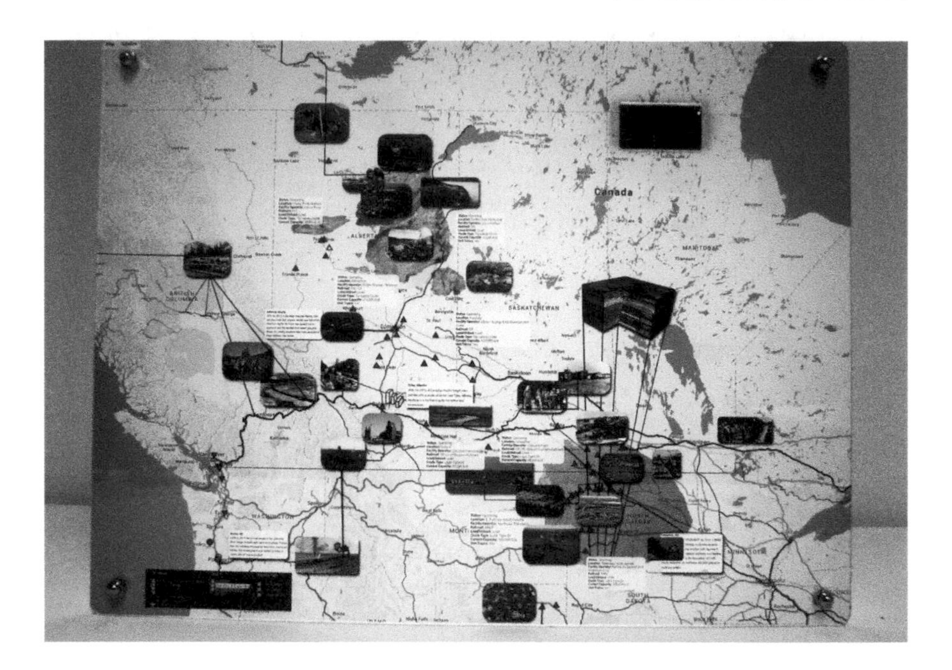

Fig. 9.4. *Mapping the Terrain: Pipelines, Bomb Trains and the Narrative of Land* (prototype), 2016.

a multitude of methods or sources (solar, nuclear, wind, natural gas, coal, and so forth), including the body. All of the human-power projects begin with the body's expenditure of energy to produce electricity, which is then consumed in and as part of an artwork, such as backlit images or videos. Two of the works mentioned, *S.OIL* and *Benzene*, reference oil specifically, while *EMERGY* does not overtly focus on oil but does allude to the use of fossil fuels (coal, natural gas, oil) and/or nuclear and hydropower forms of electricity production, as evidenced by the images of Phoenix and Las Vegas that are powered by power plants far from sight. Oil is not generally used to generate electricity, but rather as fuel for heating and for transport, including boats, trains, trucks, ships, planes, and so forth. These projects allude to the extraction, processing, and combustion of fossil fuels in order to have electricity or enable food production. *Benzene* was the first time I made the connection between fossil fuels and environmental and human health—that is, between the production and combustion of hydrocarbons and their toxic effects on ecosystems and bodies.

AZ: Along with these recurrent themes, there is also a sustained participatory and embodied dimension to your art practice. Your audience is invited to be

Fig. 9.5. *S.OIL*, 2012. Detail of planters and irrigation system.

an active part of the project (an actant). Can you talk more about the kind of choreography of experience that supports your work?

MM: This choreography, as you call it, was a deliberate choice and was inspired by my own backcountry travel experiences. I witnessed resource extraction on a massive scale while out enjoying the mountains. It was the first time I had come face to face with my own complicity with such activities. It was more than an intellectual understanding—it was physical and very visceral for me. The idea for using human power as a participatory element came while chatting with a climbing partner about the use of bicycles to generate electricity. I wanted to convey the connection between our everyday consumer demands and the resource extraction required to meet those demands. How do I get people to understand it through a physical embodiment, the way I did, without having to go climb a mountain and see it for themselves? Not that that would necessarily have a large enough effect to induce a change in behavior, though it might precipitate an awakening.

Human power has a varied history, and the use of the human body to generate power—for example, for plowing or powering a mill, as prisoner labor was used in England in the 1800s—is more of a novelty today. Although modernization has reduced the use of human power around the world, it is still used as mechanical power for tooling in some villages in Africa, India, Asia, and South America. Nonetheless, I found it to be a profound way to understand energy expenditure and production in relation to energy consumption simultaneously. I also wanted to use human power as an approach that required community effort and a certain mindfulness to achieve a state of flow in order to make the operation of the mechanism as efficient as possible. This parallels industrial machining systems: we need to make our industrial processes more efficient to eliminate waste both in terms of energy and material. When you think of it in economic terms, there are multiple implications: human labor was what we had to trade. At the same time, machines have always been like prosthetic limbs, enabling us to be more powerful in exploiting the Earth's resources. But how do we negotiate the value of our labor if our bodies are no longer needed because of machine automation? On the large-scale farm, for example, a sole farmer can combine the entire crop all on his own, without the need for more than a few farm hands. My projects brought that connection with machines, which we've historically had, to the forefront as part of a staged and "choreographed" reality.

AZ: In June 2016 you attended the BRiC residency. It was there that you commenced the new work, *Mapping the Terrain*. You have described the project as arising from an interest in and concern with energy infrastructures, specifically cross-border oil transport. At the same time, you have remarked on how the intention of the work is to transcribe the interwoven narratives of the communities affected by oil transport. Can you discuss the early stages of this new work? How does one go about suturing community narratives? And how do you conceive of your role as storyteller?

MM: *Mapping the Terrain: Pipelines, Bomb Trains, and the Narrative of Land* (begun in 2016 and ongoing) is a way to look more specifically at all the actors in the oil industry and its broader implications. The research begun at Banff was not just an exploration of the networked technologies I wanted to use as aids, but also a way of thinking about and looking more closely at energy infrastructures and the communities affected by them. I began using mapping tools to make visible the start- and end-points of the flow of oil, and the number of barrels per day each transfer point processes, mapped across the Prairies, the Dakotas, all the way east to the coastal points where the oil is transferred from train to barge. The object of *Mapping the Terrain* (see fig. 9.4) is not necessarily indicative of what the project will look like, but rather a blueprint of ideas that are currently being developed as I work with a

Fig. 9.6. *EMERGY*, 2008. Detail: rower/performer generates electricity to power the Salt River images on canyon-shaped walls for another visitor to experience an activated installation.

community in southeastern Saskatchewan, where Bakken crude is extracted, and a community in Albany, New York, where the train brings the oil for ship transport to refineries. Seeing the intricacy and density of the network of pipelines and rail lines carrying crude oil throughout Canada and the United States was mind boggling. It was so dense that I had to actually separate pipelines from the rail lines. Which begs the question, why do we need more pipelines? When viewed as a topography, borders don't mean much. But, of course, we know there are many layers to that narrative. And this is what I'm after. What are the impacts to human and nonhuman communities by these infrastructures? What are their political implications, economic myths and realities, and contentions? These infrastructures connect diverse communities from start to end, from one country to another, one type of landscape and land use to another. I am interested in revealing these narratives by mapping them through a multilayer approach—both materially and metaphorically—that includes references to human and geological history.

While the human-power projects were structured around gallery-based participation by visitors, participation in this project is more direct and is based on dialogue and co-creative object-making. The "interaction" takes place outside the realm of the gallery but will return to a gallery where the objects and documentation of the process become "framed." Therefore, the new work is a long process of durational engagement and the exchange of stories, with the goal of creating a work that's about the group's local environmental concerns regarding oil.

Two communities that I will be working with over the next few years are cross-border and very different. In southeastern Saskatchewan, an area that has historically experienced the boom and bust of fossil fuel extraction and farming, I hope to engage with several different groups, some of which benefit from oil fracking activities directly or indirectly, willingly or unwillingly, and some of which do not. At the other end of the crude oil rail line (and soon pipeline), in the neighborhood of South End, Albany, New York, lives a marginalized, low-income, and primarily African American community—people who do not benefit at all from crude oil trains in their backyard, but who bear the risks and impact to their well-being. By engaging directly with communities, I can gauge what narratives and attitudes emerge when discussing oil production, its environmental impact, and risks. I'm searching to see if there are commonalities between the narratives of the communities and how might I leverage this to foster understanding between the two through art-making. I see my role as a gatherer, collecting the stories, as well as co-creator in how those stories take visual shape. In the gallery space, I will weave the individual stories—from community members but also through my own investigations into the backstories of oil regimes—to create a larger narrative.

AZ: Your projects demand from participants a radical disengagement with the inadequacies of our energy cultures and the toll these energy regimes have on bodies, communities, ecological processes, and nonhuman life forms. Your work foregrounds this argument by rendering visible or materializing energy, both conceptually and physically. By "disengagement," I'm referencing the ways in which your work reveals the substructures of energy infrastructures, and the attempts made to render visible what is otherwise subsumed by a much larger energy system. Perhaps we might start, then, to better visualize the ways in which oil isn't necessarily something out there or contained within patches of field or sand. Rather, it is everywhere: it is culture, economy, politics, and bodies. I see this point of disengagement as a moment of realization when one comes to understand one's own complicity in petroculture.

MM: I hadn't thought of it as a "disengagement" before, but I think you may be right. While the operation of a mechanism requires a mindfulness and engagement with the physical object, durational repetitive activity can lead to a state commonly known in athletics as "the zone," a state of flow. The disengagement is with the contexts presented in the whole of the installation. But it also is a state of openness and, as you say, heightened consciousness that hopefully is a moment of realization.

One of the things I try to convey in the human-power projects, as well as some of the current work, are systems—ecological, industrial, and economic—that affect all areas of civic life. I conceptualize projects through a systems-thinking approach and consider how these systems can be exposed—not just made visible but shown to be interconnected and connected to everyday life. For example, the installation *S.OIL*, being the most complex of the human-power works, integrated several of these systems—mechanical, biological, electronic, economic—and placed the human body at the "helm." Without the complicit and explicit participation of the human body, none of these systems would activate. Perhaps the plants might survive for a while without water, but not for long. Therefore, all of these systems are created, implemented, and maintained by the human hand. Ecological systems, in a sense, don't need our input, but since we are part of the Earth's living systems, we obviously affect them.

By making these systems discernible, I ask participants what they see and what they are comfortable with, what they want to take responsibility for, and what they want to abdicate. It was interesting to observe how people interacted with the mechanisms, the source where the electricity is generated. Some took to them aggressively, shaking the machines and generating more power than was needed. Others were much more awkward or timid, lest they break the sculpture. Almost all users reacted to the feedback loop when they realized their actions made something happen. Once they realized that it was because of their labor that the videos would play or the water would get pumped, they were delighted, but they realized what hard work it was. In one case, during one of the *EMERGY* exhibitions, one visitor helped another get in the boat and explained the rowing mechanism. It was clear that people wanted to generate the electricity so others could enjoy the full spectrum of the artwork.

With each of the works I include a statement that mentions the processes of materials involved in such large installations. "Emergy," in fact, is a contraction of the words "embodied energy"—how much energy it takes to produce all the materials used in this installation. It isn't as obvious looking at, for example, a nine-foot boat and knowing that it was made with broken-up furniture that was remilled, making demands on labor instead of new resources. But one could see that the rowing machine was old and the bicycle parts used. I suppose one could choose to disengage with these realities. But in order to experience the art, you must activate it through your own volition. Each person who rows the boat, pumps the handcar, pedals the car, or offers up their story on a map becomes imbricated in those layers of meaning as an actor or performer in the work. They are performing for the viewer—cooperating, in a sense—offering up something physically or emotionally woven within the contexts of an artwork.

AZ: If I were to outline key concerns across your body of work, then I might suggest that your work attempts to render visible energy regimes as "matter" in two particular ways. First, I mean "matter" as in a physical substance that has inertia and occupies physical space. It's matter as in material. I see your work as attempting to materialize facets of energy regimes or infrastructures by revealing the actual physical components or subsystems required to power these systems and in ways that demonstrate how humans are a constituent part. Second, I mean "matter" as being of importance or having significance. In this sense, your work endeavors to demonstrate not only the physical manifestations of energy regimes, but also the ways in which these regimes are of consequence in our daily lives. You ask your audience and participants to consider why and how energy is produced, consumed, and depleted, but you also ask them to consider energy as an ethical, philosophical, political, economic, and cultural orientation. It is something that powers our homes, workplaces, or modes of transportation, but it also has the capacity to empower and disempower communities, bodies, and worldviews. What do you think about the possibility of such a twofold meaning of "matter" running through your work?

MM: When we think of energy, we do not readily think of it as material, something tactile and visible, such as a physical object. Rather, energy is abstract: that complex and illusive nonthing that is neither created nor destroyed. But because material is inseparable from mass, and on a basic level—since energy and mass are interchangeable—they are essentially different forms of the same thing. If we think of the sun's energy stored in biomass such as a tree, which then becomes heat-generating firewood, or fossil fuels such as petroleum and coal used within a system to generate other forms of energy—say, heat and steam for electricity—there is an interactive relationship between objects (or bodies) and forces that act upon them. These forms of material energy are at the core of our modern civilization. Using human-operated mechanisms, the body performs the work, and that work is measured as energy. Conversely, the body requires fuel (calories) that is transformed into energy to perform the work. Therefore, we are—our bodies are, as you say—a constituent part of the entire system. Energy as material is consumed by and expended from the body to the entire art installation.

The political rests as much in what you do see—material manifestations of energy—as what you don't see—the material manifestations called waste. What is waste? Is it the byproducts of all this materiality? Where does it hide? What problems does it cause? Who is affected? Who is responsible? Take, for example, *S.OIL*. Genetically modified corn now finds its way into our food, utensils, and fuels, to name a few. It has changed our agricultural systems, our bodies, and our energy use. But they require enormous energy inputs in the form of fertilizers, pesticides, and herbicides made of petrochemicals, which in turn require fossil fuels as their material base and energy source. The chain of waste flows down rivers and into soil and air that pollute communities of human and nonhuman inhabitants. These are political matters, especially when they enter the discourse of climate change. Although I don't consider my projects overtly political, but instead let the questions point to personal choices, those choices more often than not lead to political stances.

It is interesting that you point out the twofold meaning of "matter." It brings to mind Bruno Latour's argument between what he calls "matters of concern" and "matters of fact." Our focus has been on attending to scientific fact and knowledge production, but perhaps it is time to attend more to matters of great concern today. In other words, it's time to take action to resolve the problems that have been leading us to climate crisis. To effect broad-scale action means these issues must matter to people. We must care about these matters, about our only viable planet. The installations are tasked with communicating some of this scientific information, and perhaps, through invested physical engagement, participants will find the courage to go further. Engagement with the contexts and the aesthetic objects may provoke an emotive response, therefore, ascribing personal meaning to these "facts." Hopefully, the takeaway for participants and viewers is that these complex issues matter very much because humans and nonhumans are interconnected and affected (ecologically, politically, culturally, economically, and so forth) by energy regimes.

CHAPTER 10

The Energy Apparatus

Am Johal

Looming over the turn to energy in the humanities is an insight about the energy content of different zones of experience, and the cultural reproduction of that content when viewed at scale. Difficult to discern, however, are the implications of "the petrocultural" for the older question of human-animal relations, sovereignty, and planetary history. Energy comes up to our technical systems through a multitude of scales, species, processes, and life forms. In concept and history, energy is an opportunity to rethink important questions for philosophy and theory: How did the human become the master of the animal world? Does the existential ecological crisis create the need for a new categorical imperative? What does it mean to include energy in the definition of our human form of life? More generally, how can we figure energy as a cultural form, and what is the relation between cultural form and *form-of-life*? While Italian political theorist Giorgio Agamben's work is not generally seen as addressing questions of energy or environment, this chapter will reposition key concepts from his work in relation to the problem addressed by this book: namely, how do we conceive of energy as both technical and social, environmental and economic, microphysical and metaphysical?

These admittedly ontological questions prompted by the question of energy will lead to the challenge posed by petroculture to method and the political: if conceptions of "progress" are demobilizing today, what kind of materialist construction can be reconstituted in light of the petrocultural and its constitution? In this time of planetary nihilism and the foreclosure of being that it intimates and gestures toward, what kind of space for collective organization and the formation of resistant political power can be conceived of, constructed, and put into motion? How can discourses from different disciplines (such as continental philosophy and energy humanities) be placed in productive dialogue with one another in relation to these questions? This will be an introductory intervention in several areas of Agamben's work that should be useful in new theoretical questions regarding energy cultures.

First, what is *form-of-life*? Agamben begins through Aristotle and Wittgenstein:

> What we call form-of-life is not defined by its relation to a praxis (*energeia*) or a work (*ergon*) but by a potential (*dynamis*) and by an inoperativity. A living being, which seeks to define itself and give itself form through its own operation, and vice versa. By contrast, there is form-of-life only where there is contemplation of a potential in a work. But in contemplation, the work is deactivated and rendered inoperative, and in this way, restored to possibility, opened to a new possible use.[1]

The juridical capture of the human, within the particular apparatus of the law, is also the loss of the possibility of controlling one's own body. But if life is common to all, as Agamben argues, it can never be a property. Life, in this theoretical framing, is a politico-philosophical concept rather than a scientific one—it is a secularized political concept. Agamben argues, through Aristotle, that humans gather together not just to live, but to live well—the essence of the political act of forming community.[2] Agamben writes: "we understand a life that can never be separated from its form, a life in which it is never possible to isolate and keep distinct something like a bare life."[3] *Form-of-life* denotes the space at the fracture between the living being and the speaking being. The relation between potential and inoperativity, and the pervasive terms of its unfolding, is a fundamental aspect of petroculture; energy culture is working out precisely this question, where "form" is the ecological and material context of energy. The energy system built up around fossil fuels is itself an apparatus. The impact of energy on *form-of-life*, as part of the affect of petro-logics, is essential to recognize.

Agamben's notion of "bare life"—that place between human and animal life—seems particularly relevant to consider in relation to the expanded notion of energy occasioned by our petrocultural present, especially as it relates to the organization of collective human life. In *The Open: Man and Animal*, Agamben unpacks the political and practical separation of humanity and animality. He identifies aspects of the human that are animal in form and parts that are specific to the human. For Agamben, the "anthropological machine" of Western modernity needs to be thoroughly interrogated to understand how the contemporary human is materially and historically constituted. Recalibrating the work of the *machine* to become inoperative is an essential part of any critical act. Agamben writes:

> It is more urgent to work on these divisions, to ask in what way—within man—has man been separated from non-man, and the animal from the human, than it is to take positions on the great issues, on so-called human rights and values. And perhaps even the most luminous sphere of our relations with the divine depends, in some way, on that darker one which separates us from the animal.[4]

The connection between energy culture and animality studies emerges. For Agamben, the political conflict that governs every other conflict is that which exists between animality and the humanity of man—the caesura between human and animal life. The process of anthropogenesis, the becoming human of the human, is a forever unfolding condition—an event that never stops happening.[5] In this logic, the human is *produced*.

But just as important as the insight made available by the concept of bare life is the zone of distinction between *form-of-life* and bare life, especially in a present such as our own—named now, by various disciplines, the Anthropocene—when that zone has become both the grounds of biological life, and an expansive threat to most forms of life (both human and nonhuman). Petroculture involves not just the ways we move people and goods around the world, or how we make consumer products, but the interimplication of natural systems and sociotechnical systems. How do we begin to capture conceptually these zones of constitution as a zone of crisis?

Regarding the pervasiveness of energy culture as a large-scale *apparatus*, we could then ask, *what kind of human does petroculture produce*? The human produces energy and is in turn produced by it. The Anthropocene is also produced by, and produces, human subjects. Agamben defines the apparatus as "anything that has in some way the capacity to capture, orient, determine, intercept, model, control, or secure the gestures, behaviors, opinions, or discourses of living beings."[6] For him, a subject emerges from the conflict between living beings and apparatuses, "the everyday hand to hand combat with apparatuses."[7] The apparatus represents the structure of the de facto capture of everyday life as such. Energy culture in its very pervasiveness and colonization of human life has evolved into an all-encompassing apparatus. The political project of moving away from fossil fuels on a grand scale has to dream beyond and around this apparatus—a move that will create a new apparatus with new problems.

Within any apparatus, the possibility of human agency has a relationship to the concept of *care of the self*. Agamben discusses this notion through Aristotle, Foucault, and medieval liturgical writings, and considers the contemporary forms such care has taken and its relationship to the body. The care of the self is reflected upon as an ethical act—the preservation of the act of being as part of the work of life and a part of being in the world. In this act is the process of overcoming the revolting presence of ourselves *to* ourselves—our disgust with our own existence and innate desire to find a reason to be. One question to consider regarding Agamben's work in critical dialogue with energy cultures is what this *care of the self* might look like today within the apparatus of energy culture and saturation of fossil fuels across cultural, economic, and political structures. And what kind of "war of position" can be waged on the ultra-hegemony of petroculture?

The ecological dimension and the crisis it identifies in its collective existential form confront the very ontology of being. Agamben speaks of Heidegger's term *hingehaltenheit*, "being-held-in-suspense," and *brachliegend*, "deactivation" or "lying fallow."[8] The lack of use of the body, the very attempt to act out the uselessness of the human, becomes a political act—it is the political act of withdrawal from the apparatus that Agamben puts into theoretical play. The ultra-hegemony of energy culture and its global social affects evokes something like Ernst Jünger's concept of *total mobilization*—the full merging of the war machine, technology, and the body. The antithesis to total mobilization is in repoliticizing the act of uselessness. Building out this idea of uselessness as a political act is essential in understanding energy culture—the use of the body (*oikeiôsis*) and, simultaneously, its capacity for uselessness, are at stake. As the relationship to the use of things potentially becomes broken by the disorientation produced by the ecological crisis, the capacity to act is called into question, as captured in the expression, "I can, but I don't." In this lack is a loss of presence in our selves, a deficiency that puts us in the particular situation of experiencing the body as something improper to itself. It denotes the phenomenon that one cannot conclude anything, but can only abandon the act itself.

In the state of exception, it is the sovereign who decides bare life. For Agamben, all living beings are in a form of life, but not all are (or not all are always) a *form-of-life*. This distinction is similar to Alain Badiou's idea that anyone can be a *being*, but not everyone is a *subject*. Agamben's call for inoperativity, through *form-of-life*, opens up the possibility to make idle the merely living being. So, we have in essence a call for a new *form-of-life*, a subject that is built around inoperativity, the labor of withdrawal from the world as it is. In a time when even the subconscious has been colonized, withdrawal is a political act and a reclamation of the body, language, the unconscious, and the conditions of bare life itself. Mohawk writer Audra Simpson writes eloquently about the productive place of the politics of refusal.

The state of exception exists between public law and political fact, between the juridical order and life. As a political concept, it is the transformation of an exceptional measure to a normalized technique of government. It identifies the Roman concept of *necessitas non legem habet*—the idea that "necessity has no law" or *iustitium*—the suspension of the law. The idea of violence is separated between law-making violence and law-preserving violence. The state of exception has inclusion in the law through its legally sanctioned right to exclusion. But the ecological question and energy culture more broadly today present us with a further question: if it is the sovereign who decides on the state of exception, then what emerges if the sovereign no longer exists within the human collective? In that sense, Michel Serres's observation in *The Natural Contract* that science overtakes the law as the place of adjudication is an important one. The decentering of human adjudication of collective life appears in a number of theoretical projects today, including the posthumanism of Rosi Braidotti and Bruno Latour, as well as the speculative philosophy of Quentin Meillassoux. We could observe that this act of decentering is also breaking up the liberal traditions of governance as the dialectic between tradition and change plays out between humans and nonhumans, and humans and more-than-humans.

The concentration of energy inhibits democracy and traditional notions of sovereignty itself—it could be described as a central feature of petroculture. As Michael Ross has pointed out, characteristics such as a *rentier* effect (use of the state to ameliorate criticism from the public through spending and investment to gloss over illiberal tendencies) and a *repressive* effect (growth of state and private security to quell social unrest) have links to the neo-authoritarian rule that is a feature of this culture.[9]

The relationship between life and energy is also something that requires greater theoretical deliberation. As Howard Caygill writes,

> the theme has not been properly explored, leaving the question of the nexus of life and energy still philosophically unresolved. Perhaps this should be recognized by eliding the two terms—life/energy—and trying philosophically to understand what the implications are of their relationship without reducing one to another. It is perhaps necessary to discover a perspective on life/energy that is neither a revival of vitalism nor an endorsement of a purely physical understanding of life.[10]

Agamben writes that the suspension of juridical capture is built into *form-of-life*: "form of life is truly poetic that, *in its own work*, contemplates its own potential to do and not do and finds peace in it . . . In this sense, form-of-life is above all the articulation of a zone of irresponsibility, in which the identities and imputations of the juridical order are suspended."[11] In aspiring to a *form-of-life*, the *being of being*, "what is at stake today is life."[12] To think some of Agamben's concepts through the history of oil and energy adds important theoretical context to discussions of energy culture. Conversely, any discussion of the apparatus today without considering petroculture in depth is working in an irreconcilable blind spot. A concept of apparatus that excludes petrocultures starts to look insufficient. The zones separating *form-of-life* and bare life without consideration of the deep time of oil become inadequate.

Change, if it is to come, requires a rupture with the existing order, even of forms of resistance and how they conjure up a relationship to energy at the scale of the state and capital. In invoking Kojève's observation that the Latin-influenced countries of Europe (Italy, France, Spain) conformed to a *douceur de vivre*—a "sweetness of living" that follows Marx's notion of the humanization of free time, in comparison to the Protestant, more capitalist aspects of northern Europe—Agamben asks what this notion might mean today: the good life, the sweetness of living.[13]

As others have argued, political forms are, ultimately, shaped by energy forms. As Agamben writes in the introduction to *Homo Sacer*, "until a completely new politics—that is, a politics no longer founded on the *exception* of bare life—is at hand, every theory and every praxis will remain imprisoned and immobile, and the 'beautiful day' of life will be given citizenship only either through blood and death or in the perfect senselessness to which the society of the spectacle condemns it."[14] Functionally we could say of the domination of fossil fuels over economic, political, and cultural spheres of the present that the deferral of a new politics rests on the nonrenewable sovereignty of oil, natural gas, and coal. We appear senselessly locked into an operative fiction in which the necessity of transition to a postpetrocultural world both emerges from and is impaired by the petrocultural present. Yet here, in this admission that we are effectively the subjects of oil is at the same time a missing term, which is the historical specificity of rule, bound as it is to the time of capital. Energy culture is the space in which the sovereignty of fossil fuels is reproduced as capitalist modernity. What energy culture asks of us now, more precisely, is, *which way do we go and what kind of human will we be and how will a response on the side of the preservation of life be organized and in what time?*[15]

Notes

1. Giorgio Agamben, *The Use of Bodies*, trans. Adam Kotsko (Stanford, CA: Stanford University Press, 2016), 247–48.
2. Agamben, *The Use of Bodies*, 196.
3. Agamben, *The Use of Bodies*, 207.
4. Giorgio Agamben, *The Open: Man and Animal* (Stanford, CA: Stanford University Press, 2003), 16.
5. Agamben, *The Use of Bodies*, 111.
6. Giorgio Agamben, *What Is an Apparatus?*, translated by David Kishik and Stefan Pedatella (Stanford, CA: Stanford University Press, 2009), 14.
7. Agamben, *What Is an Apparatus?*, 19.
8. Agamben, *The Open*, 66.
9. Michael Ross, "Does Oil Hinder Democracy?" *World Politics* 53, no. 3 (2001): 325–61. Increasingly, this phenomenon and particular aspect of energy culture could be called *neo-authoritarian extractivism*.
10. Howard Caygill, "Life and Energy," *Theory, Culture & Society* 24, no. 6 (2007): 19–27.
11. Agamben, *The Use of Bodies*, 247–48.
12. Agamben, *The Use of Bodies*, 209.
13. See Silvia Mazzini, "On Giorgio Agamben, the 'Latin Empire' and a European Identity," *Aljazeera*, 16 May 2014, https://www.aljazeera.com/indepth/opinion/2014/05/giorgio-agamben-latin-empire-e-2014516133117453267.html.
14. Giorgio Agamben, *Homo Sacer: Sovereign Powers and Bare Life*, translated by Daniel Heller-Roazen (Stanford, CA: Stanford University Press, 1995), 11.
15. Giorgio Agamben, *Infancy and History: On the Destruction of Experience*, translated by Liz Heron (London: Verso Books, 1993), 99. "Every conception of history is invariably accompanied by a certain experience of time which is implicit in it, conditions it, and thereby has to be elucidated. Similarly, every culture is first and foremost a particular experience of time, and no new culture is possible without an alteration in this experience. The original task of a genuine revolution, therefore, is never merely 'to change the world,' but also—and above all—to 'change time.'"

CHAPTER 11

Aeolian Survey

Hannah Imlach and Thomas Butler

Standing in front of the large map, you can choose what to listen to. Over thirty sites across the city of Glasgow, Scotland, are depicted: from the Victorian rooftops around central George Square, the elevation of Queen's Park in the south, and the summits of high-rises throughout the city, the sound rushes in. Quiet at first—a soft cluster of low buzzes fading in and out—soon the space will be occupied by billows of held tones, each growing and blooming asynchronously in a smear of heavy-metal distortion. And then, as swiftly as it arrived, the gust of sound peters into a peal of bare sine waves, stored shadows of each pitch we have just heard. Throughout, a staccato counterpoint of familiar bleeps, clicks, and whirs has forced an interplay between the abstract and the everyday: at the peak of each distorted swell we hear echoes of our electronic lives—microwave oven rumble, skeuomorph ringtone, aircon hum, canned sitcom laughter . . .

This is *Aeolian Survey*, a hypothetical installation that we have developed. It comprises a network of sculptural aeolian harps installed in a cityscape. As they are played by the wind, their heavily processed sound is transmitted to a gallery listening space where members of the public can eavesdrop on the renewable energy potential of a city-center site. You can listen from home, too—an interactive web app livestreams the sounds direct to your computer or smartphone. This allows you to easily check in on the installation during varying weather conditions, but the experience is not as immersive as being in the gallery space, where large speakers retell the vibrations of the harps' strings in visceral and imposing detail.

The harps themselves sit placidly among the old chimneys, new ducts, and vents that litter the roofs above Glasgow's commercial districts. They are imposing structures more than six feet tall, an amalgam of triangular forms arranged to suit their immediate environment: some seem to stay low, clinging to the rooftops, while others stand tall. Strung with steel cabling, they are equipped with electric guitar–style pickups that convert the energy imparted to the strings by the wind into an electrical signal. When installed in multiple locations, they act as a synaesthetic mapping device, conveying the potential of local wind power through sound alone. To their audience, the harps make strange our everyday energy culture—an aural figuration of energy—and promulgate its infrastructure through aestheticization.

Their construction from white-painted metal is reminiscent of the many wind farms constructed since the 1990s in the Scottish landscape and offshore. These arrays of tall turbines are now a common sight here, emblems of the nation's slowly changing power generation priorities and policies. They are also aesthetic objects in their own right: some see them as eyesores, ugly intrusions into the "natural" countryside. They could also be viewed as giant kinetic sculptural forms that help liberate the country from fossil fuel dependency, a necessary step toward creating a future sustainable Scotland through utilizing the country's massive potential for harnessing renewable energy sources.

In 2015, Scotland produced the equivalent of 59 percent of its electricity consumption though renewables, predominantly wind power.[1] For now, the Scottish government's stated aim of generating 50 percent of *all* energy expenditure though renewable energy technologies by 2030 seems remote.

Fig. 11.1. Proposed site for *Aeolian Survey I*, Glasgow city center. 3D model and digital photograph, 2017.

But a successful, fossil-free, environmentally and economically sustainable electricity generation system has already been installed in Scotland, albeit in miniature: since the 1997 community buyout of the Isle of Eigg (which lies ten miles off the west coast of Scotland), a pioneering renewable energy grid has been installed on the island and maintained by its community. Comprising solar, hydro, and wind power, Eigg's renewable infrastructure is an extraordinary achievement and demonstrative of the benefits of relocalizing energy production, an idea that has had a significant influence on the work we present here. However, the Isle of Eigg would have little need for our *Aeolian Survey*, as parts of its energy production are inherently sonified: a resident remarked to us that heavy rainfall increases the sound levels of the waterfall at Cleadale, signifying an increased yield of domestic power from the hydroelectric turbine.

Transposing Eigg's electricity generation strategy to the mainland and re-creating it on a nationwide scale is unfeasible (the island has a rare combination of low population density, thus a relatively small demand for electricity, and an ideal source of hydroelectric power). Nevertheless, it is the relationship between autonomy and energy that is most compelling about the situation in Eigg. Since

Fig. 11.2. Proposed site for Aeolian Survey II, Glasgow city center. 3D model and digital photograph, 2017.

the island is unconnected to the national grid, before 1997 the residents relied on noisy and polluting diesel generators for their electricity: independence from a neglectful absentee landlord curtailed this necessary reliance on fossil fuels.

In Scotland more widely, self-determination and petroculture are thoroughly intertwined, and the 2014 referendum on whether Scotland should leave the United Kingdom and become, once again, an independent nation was colored by crude North Sea oil. Against a backdrop of volatile, tumbling oil prices, pro-UK forces regularly insisted that Alex Salmond, the then first minister of Scotland, leader of the Scottish National Party, and figurehead of the Scottish independence movement, had overstated the future benefits of North Sea oil to the Scottish economy. Relying on it would allegedly compromise the quality of life for residents in a postseparation Scotland. Meanwhile, pro-independence campaigners looked to Norway, our oil-rich neighbor, as an exemplar of small-nation petroprosperity. Scotland may be edging toward a sustainable, renewables-based energy system, but in the national psyche at least, quality of life, economic well-being, future prosperity, and confidence in self-determination seem inextricably linked to the black stuff.

Fig. 11.3. Proposed site for *Aeolian Survey III*, Glasgow city center. 3D model and digital photograph, 2017.

At the time of writing, the Scottish Parliament is preparing to debate the prospect of a second independence referendum. With the United Kingdom having chosen to leave the European Union, there will be many threads for voters to untangle if the plebiscite does indeed go ahead. However, it is likely that speculation about the future of energy will play its part, even on a metaphorical level (although it is worth noting that the current Conservative UK Government has severely depleted investment in the renewable energy sector).

As artists, we cannot offer real-life solutions to real-life energy issues. Instead, we consider *Aeolian Survey* to be a semi-speculative fiction, a playful space where alternative energy futures can be imaginatively conceived. Mindful of the present, we mix the roar of the electrified harps with location recordings made inside the very buildings they crown, a sonic cross-section that collates evidence of electronically enhanced human activity—everything from heart monitors to toasters, boilers to laptops, strip lighting to e-cigarettes. Thus, we sonify both the energy resource and its expenditure.

The maelstrom of overdriven harp sounds is now beginning to ebb, and the downpour of computer keyboard clicks slowly fades away. A moment's silence. Soon the major-chord sine waves will return, crescendoing from nothing, telling us that the building's storage batteries have come back online.

Notes

1. Scottish Government, *Energy Statistics Summary—June 2017*, http://www.gov.scot/Resource/0052/00521900.pdf.

CHAPTER 12

Anecdotal Encounters on Driveways: The Aesthetics of Oil in Northern Alberta and Newfoundland

Megan Green

Arising from an encounter with an errant deer's leg on a Fort McMurray driveway, *What exactly were we supposed to have learned from the fire?* (2016; fig. 12.1) draws on personal anecdote to reconsider the aesthetics and narratives of oil. It reconceptualizes the landscape where the leg was found as a space of new encounters with the geocultural landscape in which oil is embodied. Through material practice and critical elaboration, I attempt to convey the humorous and melancholic conflation of the strangeness and banality of a place considered by many to be dystopic and somehow otherworldly.[1] My work explores industrial interactions with the landscape and the manner in which these interactions are tied up in narratives of cultural reproduction and class. It seeks to complicate the petropoetical[2] aesthetics of oil by communicating the affect and cultural logics of people in regions associated with the energy industry: Newfoundland and Alberta. These regions are linked not only as sites of oil extraction in Canada, but also by the movement of people from east to west in an economic migration between the island of Newfoundland and Fort McMurray in northern Alberta. As I am a Newfoundlander who was a part of this migration, my work considers being at home in "the offshore," so called to denote remote spaces of oil extraction, and being female in a space often rendered in macho terms. Many narratives are being told about oil, while others remain to be told—the sense and sensibility of living in and around extraction sites as a classed, gendered body. *Untitled* (2013; fig. 12.2) is a medical bottle for male urine containing a "feminine," small furry object. The object does not appear to fit easily through the opening of the bottle. The work is intended to express an awkward humor and melancholy about a culturally loaded feminine body degraded, limited, and disturbed by a patriarchal narrative. In the context of Canadian colonialism and its geopolitical spaces of oil extraction, it might also be read in relation to colonial dynamics. My artwork tries to visualize the affect of *being in relation to oil* and the cultural narratives surrounding it. The sometimes nonidentical relationship between the affective landscape and visual representations of these places is a focus, as is the relationship of my work to the notion of a "Newfoundland Diaspora." Material culture connoting the presence of Newfoundlanders in Fort McMurray is ubiquitous, "visible on approximately 90% of the community's streets" surveyed in Fort McMurray during a 2009 research period.[3] Generally, this material culture implies a sort of nationalistic cohesion among Newfoundlanders. Indeed, the city possesses Canada's largest population of mainland Newfoundlanders,[4] and so in discussions of Fort McMurray and petroculture in Canada, the migration of this population is a common topic of interest.

Fig. 12.1. *What exactly were we supposed to have learned from the fire?*, 2016. Nails, rope, candle, wood, shoe sizer, gold mirrored plexi. Size variable.

One of the oft-commented issues about oil extraction is that it takes place at a distance from the spaces in which most of the consumers of the resource live. But there are some for whom there is no such resource gap—those like myself who have grown up in and around the resource, for whom the landscape of oil is the landscape itself. We didn't realize that there was anything exceptional about our lives until journalists, academics, and environmental activists came and told us there was. The abnormal is here normalized, apparently. We've fast become accustomed to an aesthetic of oil and resource extraction whose aim is to shock, and which assigns to those who live near extraction sites culpability in environmental destruction and denigration that those living far away supposedly do not share.[5] My work explores another aesthetic: oil written into the banality of everyday domestic life, oil as always already, in embodiment, *the same as* that life, and so not capable of being rendered at a distance and as far from thought as from the sight of those at an ostensible distance from its extraction. Oil as multifaceted, "the fetishized ur-commodity of modern global capitalism,"[6] enabler of contemporary North American life and a meaningful complication in interactions with a physical and narrative space that is, in northern Alberta, defined by interactions with both wilderness and an extremely industrialized landscape. I feel that my sense of normalcy about the oil sands is a result of a cultural trajectory and that the assumption that it has been "normalized" obscures something about our intertwined yet precarious relationship with natural resource use and fails to describe how these things can be "geo-culturally uneven."[7] It is my feeling that approaching these issues through nuanced perspective-taking would help facilitate grappling with energy issues and climate change, and that this would include a critique of some implicit narratives present in the environmental humanities.

Feral Suburbia

In describing the built quality of Fort McMurray in "Feral Suburbs: Cultural Topologies of Social Reproduction, Fort McMurray, Canada," Rob Shields writes:

> Fort McMurray's suburbs are arguably well-planned housing developments yet inadequate, unsatisfying places to live in the sense that lack of services and consumption opportunities is compounded by a sense of geographic distance and isolation . . . Feral also suggests "savage" suburbs . . . impenetrably boggy forest at the end of the street, which exists in striking contrast to the vinyl-clad exteriors of suburban houses and their vehicle- and RV-filled driveways and streets which could be almost anywhere . . . Suburban life in Fort McMurray means to be melancholically governed

Fig. 12.2. *Untitled*, 2013. Medical bottle for male urine, fur, doily. 13 × 5 × 4.5 inches.

> most of the time by the isolation imposed by the expanse of wilderness, to be fatally forced to respect its climatic extremes, to grimly wrest oil wealth from the strata of the area, or to joyfully respond to its beauties . . . It conforms to images and forms of the North American middle class rather than to those of the local, the working class or to the village rubric of the surrounding Métis, Cree and Dene settlements. Indeed this is a paradise of the lumpen proletariat, wealthy on overtime and double shifts.[8]

Are we in Fort McMurrary modeling some quality of the pleasant veneer of modern suburban life because many of us are newly middle class? A group of people who want to be on the way up, modeling whatever concept of "up" might have been dangled before us? Is there also a judgment about what can be considered 'authentic' and local to the place considered in this text? Might this text appear to Newfoundlanders as reinforcing stereotypes "portray[ing] the Newfoundlander as a 'barely civilised' half-brute who had been beaten into submission by . . . centuries of neglect, oppression, and ceaseless grinding poverty,"[9] as white-trash "Mexicans with sweaters"?[10] For Newfoundlanders, how might this text unintentionally demonstrate how, "as patterns of out-migration follow provincial economic disparities, class hierarchies frequently become mapped onto regional and cultural identities"?[11]

As noted above, *What exactly were we supposed to have learned from the fire?* arose from an encounter with an errant deer's leg on a Fort McMurray driveway in one of Shields's "feral suburbs," and emerged while I was reflecting on this encounter during the evacuation following the Horse River Wildfire in spring 2016.

Kitsch

Can these places have their own quality? Can the ramshackle plastic impermanence of trailers and McMansions in this wild context offer a unique way of being and living? What sort of encounters might these spaces engender, and what new might be seen in them?[12] How do our "stories"[13] interact to produce this place and its relevance to the Canadian narrative? Kitsch—petroleum plastic objects, vinyl siding, and cheap home decor—and its association with class is used in my pieces with an awareness of the stereotypes associated with these regions. Kitsch is a tool for getting at the problem of the embeddedness of energy in our material environment. Graeme Macdonald writes that "the short-lived era where oil was almost universally celebrated as an emancipating, 'good' substance has long receded,"[14] yet this optimism persists in our material realities. The kitsch that I employ draws attention to "particularly embedded kinds of energy" that participate in "organizing and enabling a prevalent mode of living, thinking, moving, dwelling and working."[15] The failures of modernism associated with kitsch[16] and suburban architectures necessitate that we ponder utopian energy futures that remain filled with glittering visions from our oil-fueled past and plastic-filled present.

Black Magic (2016; fig. 12.4), *Untitled* (2013; fig. 12.5), and *Untitled* (2016; fig. 12.6), utilize plastic to imply a material embodiment. *Untitled* (2013) mixes kitsch objects that are implying a transition to a fantasy of the wild through the animal, and actual antlers cut to recall logging. *Black Magic* and *Untitled* (2016) reference an interaction of the "toxic sublime,"[17] an aesthetic generally associated with images of massive industrial landscapes, and Fort McMurray specifically, with the alternatively wild landscape of northern Alberta. While these objects are in some senses an exaggeration, however imaginative we might be, we will continue to have to work with a landscape filled with oil objects and infrastructures, complicating conceptions of the Canadian rural landscape.

Home Décor (2014; fig. 12.7) references kitsch taxidermy and features an image of a Fort McMurray resident holding two severed deer heads. This piece originated with a request I made to a local hunter for a set of antlers. Following hunting season, I was sent a cellphone image, and the antlers arrived by mail in

Fig. 12.3. *What exactly were we supposed to have learned from the fire?*, 2016. Detail.

a comically oversized box, with some fur still attached. The object includes a found wood carving, is mounted on faux wood paneling and lights up, but does so quasi-mystically with its power cord dangling. The figure appears to have emerged from the darkness onto a driveway. The antlers are not large and do not have an unusual abundance of points; they are not valuable as a trophy, and therefore fall into the realm of kitsch taxidermy.[18] This is a melancholic artifact. Rachel Poliquin writes:

> In the North American imagination, secondhand hunting trophies have long been synonymous with all things hick, kitsch, or tongue in cheek. Dusty heads and antlers mounted in restaurants, gas stations, and bowling alleys are as much a part of backwater American landscape as plastic flamingos and velvet paintings of Elvis. The heads offer a peculiar aura of failure and ruin.[19]

The image in *Home Décor* is meant to provoke a response. The man in the image is doing something that reads in particular ways to different audiences: for some, an image of a strange man doing stranger things in a northern locale;[20] for others, a person finding deer meat to eat. The man in this

Fig. 12.4 (*top*). *Black Magic*, 2016. Plastic, gold-mirrored acrylic, fur, lightbox. 11.5 × 10.5 × 4.5 inches.

Fig. 12.5 (*bottom*). *Untitled*, 2013. Antler, found plastic "animals."

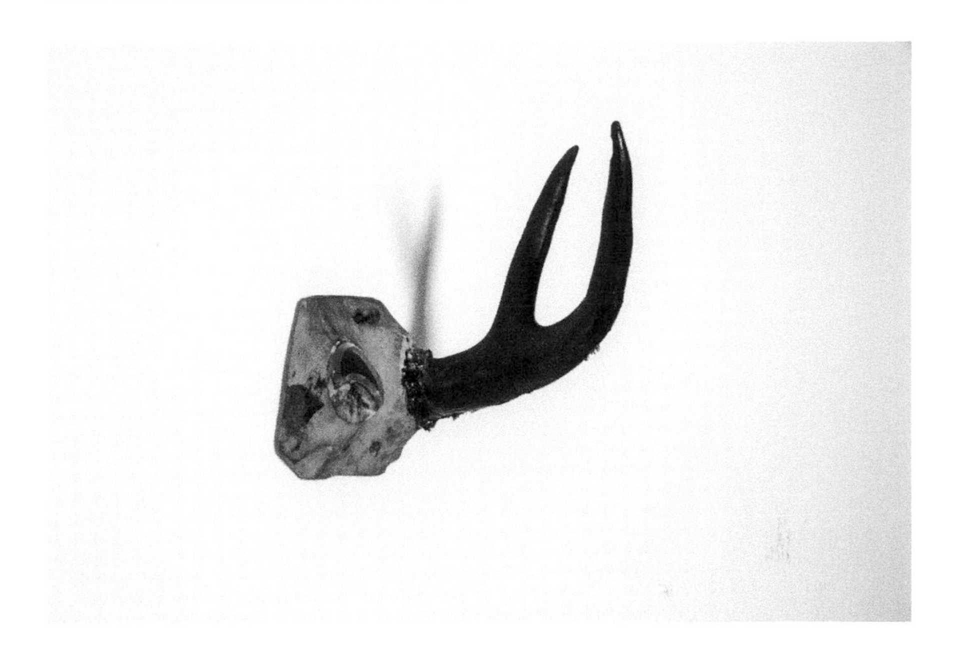

Fig. 12.6. *Untitled*, 2016. Gold plexi, black plastic, wooden antler mount.

piece is both a hunter and a worker in the oil sands. He is wearing an orange coat in a hue of safety in industry and in hunting, a bright color to prevent being killed on the job or accidentally shot. *Home Décor* is an articulation of the bind of working in a natural resource industry: as is the case for many workers, the figure holds two heads—the two minds with which one inevitably navigates the spaces of resource extraction. It is a contextually offensive image, or alternatively, a wry marker of machismo and accomplishment. The hunter in this image is performing something with an irreverent humor—something seemingly unrelated to oil, but, of course, oil is still present. This engagement with "nature" through hunting points to another form of natural resource extraction. The hybridity in this image illustrates that, although these things should not sit well together—as Poliquin writes of taxidermy, "ethics and camp do not sit well together"[21]—yet they are together. This gets at something of the complex conditions of a local subculture wherein camp and ethics have a nuanced and varied relationship—the man in the image might be doing something some would find distasteful, but he is not being a hypocrite about it.

Fig. 12.7. *Home Décor*, 2014. Found cellphone photo, lightbox, electronics, found wood carving, antlers, imitation wood paneling. 21 × 13.5 × 11 inches.

A "Newfoundland" Point of View

Oil extraction, like other interactions with the landscape, is reconciled and reproduced through narrative. Newfoundland has a cultural narrative that associates labor interactions on the landscape with a sense of belonging to place. It is a narrative that exposes a rationality that I have come to see as unconsciously informing many of my own attitudes about situation in the oil sands over time.

Newfoundland's coveted and mythologized out-port life and other nostalgic folksy narrative leanings are embedded in its regional identity, expressed through tropes appearing over and over in our cultural texts,[22] including literature, contemporary art, and the sanitized and color-saturated images that dominate recent Newfoundland and Labrador tourism commercials. Although out-port life is generally associated not with oil and lucrative long hours but with fish and precarity, it is economically maintained by offshore oil drilling and work in the oil sands.[23] A Newfoundlander might see oil as what holds off a repetition of government-mobilized resettlements of communities in remote locales around the island and of the out-migration of the past half-century. In an important sense, oil keeps Newfoundland close to itself.

Newfoundland artist Pam Hall's recent work is an encyclopedia of knowledge integral to relationships of land and labor in Newfoundland.[24] Hall is attempting to conserve traditional practices and position that knowledge as situated "in living bodies in particular places."[25] In doing so, she is challenging maligning stereotypes by revaluing the "knowledge" of people from this place. Hall positions these forms of knowledge in conversation with other forms of knowledge that tend to erase "the qualitative, the embodied, the value-laden, and many individual and cultural ways of knowing that form and inform our embedded relationship within our now endangered eco-systems."[26] In *Towards an Encyclopedia of Local Knowledge*, Hall catalogues fishing and hunting practices indispensable to local culture and to the ability of locals to survive physically and thrive culturally, and, in my view, to conceptualize the conditions of their interactions with the landscape and domestic spaces. I extend this integral cultural imperative to the role of mining and migratory work in the culture of Newfoundlanders.

The practice of mining, fishing, and sealing, and other interactions with natural resources, are so embodied in the life of the local as to make it almost indispensable in the narratives of this place. *Newfies in the North* (2014, fig. 12.8) comically refers to labor with a work sock, and an economic situation in Newfoundland that tends to leave us "gutted" by a mismanagement of, lack of control over, and reliance upon precarious natural resources. But it also refers to an amateur instructional hunting video of similar title, *Newfies in the Yukon*, available for sale at a "Newfie" diner in Fort McMurray circa 2012.[27] Newfoundland presents an interesting example of conflict and contestation between popular perceptions, cultural narrative, industry, and nationalism. Stereotypes about Newfoundland are perceived as being perpetuated by "mainlanders," and yet, as Jennifer Delisle writes, there is also a "desire for the exotic traces of Newfoundland culture [which] raises serious questions about cultural authenticity, appropriation, and the line between the preservation of culture and the perpetuation of regional stereotypes. If literature shapes conceptions of Newfoundland identity, what is that identity? Who has the right to construct it? [Townies? People from out-ports?] What about Diasporic Newfoundlanders? . . . Claims to cultural authenticity are asserted and denied over time."[28]

An ethos running through "Newfoundland" cultural texts positions the "Newfoundlander" of European origin, the "livyer," as somehow of the land and so diasporic anywhere else. It is almost fatalistically as though, through our work on the land, our interactions with the elements, the ocean, the boreal

Fig. 12.8. *Newfies in the North*, 2014. Antler, sock, imitation wood paneling. 8 × 12 × 12 inches.

forest, and oil, and our ruin in these pursuits, we may be generating a fantasy that we become "of the place" in some essential sense.

Terry Goldie identifies a strain in the cultural narrative that assumes an absence of the indigene in Newfoundland owing to a collective sense that the Beothuk likely no longer exist and a dispute as to whether or not the Mi'kmaq, "resident in Newfoundland at least since the seventeenth century, qualify as indigenous to the island."[29] However, he points out that "the assumption that the Beothuk were the only Indigenous peoples of the island seems untenable,"[30] and I would note that the Miawpukek First Nation positions its people as having occupied the island prior to this, and refer to the formation of the Qalipu First Nation. I could easily be seen as a colonial actor bringing with me, to northern Alberta, the baggage of a group with an extremely particular and often troubled relationship to indigenous issues, practically and in some of our poetic conceptualizations of ourselves as Newfoundlanders. But as Goldie writes, "the Newfoundlander is not just the person of the land, but the person of Newfoundland [as in, the province and failed nations as opposed to the island]—and perhaps the person with [a] story of the land," the land coming

Fig. 12.9. Ernst Logar in Fort McMurray, July 2016.

to define us, and not us it.[31] But what does this mean for my work, with the image of white male subjectivity I am invoking for contemplation, emerging from the darkness to grin into a cellphone, severed heads in hand?

Whether the "Newfoundland story" is defensible or not, what does it mean for the discussion about energy, which often relates to negotiating indigenous land ownership and nationalistic interests? Should the Newfoundland story's cultural thread be taken on, in all its precious exceptionalism, or is that inappropriate? What might this mean for our understanding of offshore oil and its relationship to the ostensible distance imagined between Canadian cultural centers and industry, if anything? Furthermore, as a Newfoundlander outside of the place from which we claim provenance, am I a colonial actor participating in a continuation of an abusive legacy in the same way that a cultural outsider might see the figure in the image in *Home Décor* to be? How do the prevailing cultural narratives shape, across multiple spheres, how Newfoundlanders relate to oil? And what might this mean for public perceptions of oil, especially in Fort McMurray, where the relationship of local indigenous groups to oil is in some ways and in some instances potentially atypical and complex?

The Fire

And then there was the big fire. While it makes easy narrative and poetic sense, the role of the oil industry itself in the fire is not causal, even if the spark that started it did originate in some extraction-related installation, or can be mapped onto the careless act of a stereotyped figure riding a quad, discarding a lit cigarette or stupidly lighting an illicit campfire too close to tinder. The title of *What exactly were we supposed to have learned from the fire?* (2016) was intended to be somewhat sarcastic; were there no oil sands mining, that would not solve the intensifying problem of energy, climate change, and capitalistic exploitation. Making the region the poster child for these issues could act as a diversion from how deeply ingrained problematic energy is in our society and how difficult and absolutely necessary it is to grapple with it.

Ernst Logar and I met in Fort McMurray and visited the Abasand Heights neighborhood where I had grown up, and which was largely burned down (fig. 12.9). It was also the neighborhood where I encountered the errant deer's leg. It was a relief for me to find that the experience did not ring of disaster tourism, or something else vaguely exploitative, but was an ambivalent encounter with both a surreal event and people who have lived through an unsettling experience. The fire had left piles of ash and pools of melted materials; a crust of white tackifier had been spayed to prevent the spread of harmful particulate. I was able to collect plastic from the melted roof and walls of a bus stop. I used the plastic in a sculpture, *Untitled* (2018), intended to convey the effect of the experiences during the fire of people I knew, and my own experience from a distance in Toronto sitting up that first night. To paraphrase one of Ernst's comments: "it's disturbing . . . all of these things melted together. It's like chaos."

Notes

What exactly were we supposed to have learned from the fire? (2016), *Home Décor* (2014), *Newfies in the North* (2014), and *Untitled* (2016) documentation images by Brian Limoyo Photography. This chapter utilizes some research and artwork included in my MFA thesis work at the University of Waterloo, completed in 2014. *Untitled* (2018) was completed with support from the Ontario Arts Council in the form of a Visual Artist's Creation Projects: Emerging Artist Grant. This chapter benefited from changes made by the editors of this volume, and feedback from several peers.

1. Megan Green, "Artist Statement," in *for the time being: 2017 Alberta Biennial of Contemporary Art*, edited by Kristy Trinier and Peta Rake, exhibition catalog (Edmonton: Art Gallery of Alberta and Walter Phillips Gallery, Banff Centre for Arts and Creativity, 2017), 66.
2. Graeme Macdonald, "Till a' the Seas Gang Dry? Petro-Littorals and Maturing Fields

North to South," conference presentation, "Petrocultures 2016: The Offshore," Memorial University, St. John's, Newfoundland, 1 September 2016.
3. Craig T. Palmer, "License Plates, Flags and Social Support Networks: The Symbolic Cultural Landscape of the Newfoundland Diaspora in Ft. McMurray, Alberta," *Material Culture* 42, no. 1 (2010): 2.
4. Geo Takach, *Will the Real Alberta Please Stand Up?* (Edmonton: University of Alberta Press, 2010), 171.
5. Imre Szeman and Maria Whiteman, "Oil Imag(e)inaries: Critical Realism and the Oil Sands," *Imaginations: Journal of Cross Cultural Image Studies* 3, no. 2 (2012): 46–67.
6. Graeme Macdonald, "Research Note: The Resources of Fiction," *Reviews in Cultural Theory* 4, no. 2 (2013): 4.
7. Macdonald, "Research Note: The Resources of Fiction," 7.
8. Rob Shields, "Feral Suburbs: Cultural Topologies of Social Reproduction, Fort McMurray, Canada," *International Journal of Cultural Studies* 15, no. 3 (2012): 1–11.
9. Robert C. Thomsen, *Nationalism in Stateless Nations: Selves and Others in Scotland and Newfoundland* (Edinburgh: Birlinn Ltd., 2010), 57.
10. Cory W. Thorne, "Mexicans with Sweaters, Working in the Oil/Tar Sands, in Newfoundland's Third Largest City: Regionalism, Narrative, and Imagination in Fort McMurray, (Texas?)," *Ethnologies* 34, nos. 1–2 (2012): 29–58.
11. Jennifer Bowering Delisle, *The Newfoundland Diaspora: Mapping the Literature of Out-Migration* (Waterloo, ON: Wilfred Laurier Press, 2013), 121.
12. Green, "Artist Statement," 66.
13. Terry Goldie, "Is *Galore* 'Our' Story?" *Journal of Canadian Studies* 46, no. 2 (Spring 2012): 85; Edward J. Chamberlin, *If This Is Your Land, Where Are Your Stories: Finding Common Ground* (Toronto: Vintage Canada, 2004).
14. Macdonald, "Research Note: The Resources of Fiction," 3.
15. Macdonald, "Research Note: The Resources of Fiction," 4.
16. Celeste Olalquiaga, *The Artificial Kingdom: A Treasury of Kitsch Experience* (New York: Pantheon Books, 1998).
17. Jennifer Peeples, "Toxic Sublime: Imaging Contaminated Landscapes," *Environmental Communication: A Journal of Nature and Culture* 5, no. 4 (2011): 373–92.
18. Megan Green, *The Jackalope in the Room: An Installation*, MFA Thesis Support Document, University of Waterloo, 2014, 7.
19. Rachel Poliquin, *The Breathless Zoo: Taxidermy and Cultures of Longing* (University Park: Pennsylvania State University Press, 2012), 165.
20. Sherrill E. Grace, *Canada and the Idea of North* (Montreal: McGill-Queen's University Press, 2002), 90–92; Shields, "Feral Suburbs: Cultural Topologies of Social Reproduction, Fort McMurray, Canada," 1.
21. Poliquin, *The Breathless Zoo: Taxidermy and Cultures of Longing*, 167.
22. Delisle, *The Newfoundland Diaspora*.
23. Sam Synard, panelist, "Cultures of Mobility in Oil Production Zones—Worker and Community Experiences and Boom and Bust," presenter-organized panel at conference "Petrocultures 2016: The Offshore," Memorial University, St. John's, Newfoundland, 2 September 2016.
24. Pam Hall, *Towards an Encyclopedia of Local Knowledge: Excerpts from Chapters I and II* (St. John's, NL: Breakwater Books and ISER Books—Faculty of Humanities and

Social Sciences Publications, Institute of Social and Economic Research, Memorial University of Newfoundland, 2017).

25. Hall, *Towards an Encyclopedia*, 19.
26. Hall, *Towards an Encyclopedia*, 23.
27. Deon Dicks and Daniel Robinson, *Newfies in the Yukon*, video (2010).
28. Delisle, *The Newfoundland Diaspora*, 5.
29. Goldie, "Is *Galore* Our Story?" 85.
30. Goldie, "Is *Galore* Our Story?" 85.
31. Goldie, "Is *Galore* Our Story?" 87.

CHAPTER 13

Energy Meets Telepathy Aesthetics and Materialist Consciousness

Jacquelene Drinkall

One has to stop and think! There is nothing mystical about the fact that ideas and words are energies, which powerfully affect the physico-chemical base of our time-binding activities.

—Alfred Korzybski

Mindreading: Some Thoughts on Energy

This chapter considers the relationship of energy to materialist consciousness, and thus to some speculative and fantastic work with brain plasticity, consciousness, mentality, neural interface, and even quantum neurobiology, in the service of aesthetic theory and art practice.[1] I will first look at energy in terms of old and new materialist philosophy to introduce energetic telepathy aesthetics. I will also build on my existing work that investigates telepathy in relation to materially engaged extended cognition, and my understanding of telepathy to be a form of energy transfer. In previous work I have shown how artistic and psychoanalytic work with telepathy is aligned with work on energy, mind-body-world connectivity, and transference.[2] In this chapter I show how my recent artwork is turning more self-consciously toward energy cultures, while engaging with neuroscience, the neurobiological sublime, and the "telepathologies of cognitive capitalism,"[3] as well as an intensified engagement with energy humanities, from Vaclav Smil and Jane Bennett to McKenzie Wark's *Molecular Red* and Catherine Malabou. I attempt to make apparent my engagement with energy cultures and fossil fuels within the context of my experimental work with Telepathic Art.[4] Fossil fuels are, of course, embedded within almost all aspects of our lives, and our work with them results in energy as well as greenhouse gases, of which nitrous oxide is by far the worst. Nitrous oxide, also known as "laughing gas" anesthetic, inspired William James's work with telepathy as conveyed in his book *The Varieties of Religious Experience*. I also wish to show the intention of my energetic work toward further divestment from fossil fuels and to find new forms of molecular liberation from accelerating nitrous oxide emissions. Some of my artworks literally document and perform with architectures that have high electrical energy needs, appropriating lumps of coal and graphs of oil importation into art experiments, while others work with more abstract visualizations of seemingly invisible and immaterial energetic forces involved in extended cognition and material engagement between brain and body, technology and world. Through this twofold aesthetic strategy, I hope to cultivate a telepathic neuroplastic aesthetics that may assist the brain's energy cultures and ethical habits in our efforts to move beyond fossil fuels. Overall, I wish to explore energy in order to better understand the forces within matter that shape transcommunicative consciousness beyond the confined comprehension of humanism.

Of course, I do not want to pioneer a mainstream broadband telepathy to replace existing data and telecommunications networks and thus reduce energy demands, although this may one day be possible, given Black Mirror futurism and actual current work with brain computer interfaces. Take, for example, SmartStent stentrodes—sensors implanted in the brain via delicate cardiovascular wires and electrodes—developed by entrepreneur and cardiovascular neurosurgeon Tom Oxley at the Defense Advanced Research Projects Agency (DARPA) and University of Melbourne. Oxley's stentrodes are closely related to many other developments in neural and brain computer interface techlepathies of DARPA, Google, Facebook F8's (Building 8) nonimplanted telepathy headsets, Elon Musk's Neuralink, and more. Instead, I hope to suggest telepathic transference as a form of independent artistic neuroplasticity away from corporate mind control and the ecologically destructive forces of capitalism. All this data mining and conflation of brain, artificial, and market intelligences could be the "great twenty-first-century mind frack,"[5] given that simulating a human brain would require most of the power of Europe: "For a human brain you would need one trillion watts to stimulate it. It would consume five times more power than the total generating capacity of Germany and still run 1,000 times slower than biology . . . To simulate a day's thinking, it would take years."[6]

Vaclav Smil's work on general energetics details the etymological roots of the word "energy" as it is first traced to ancient Greek knowledge of Aristotle.[7] Aristotle's *Metaphysics* defines *energy* as "actuality" and "complete reality," resonant with kinetics, work, and *entelechia*—and this concept of *entelechia* is particularly useful in connecting energy, telepathy, and materialism. Via nineteenth-century science and technologies of energy, we have a general knowledge of energy in relation to fuels, radiation, electricity, metabolism, photosynthesis, and fossil fuels. Renaissance thought on energy may seem more exotic today than Aristotle's: Galileo Galilei thought heat was a sensory illusion of mental alchemies, and Francis Bacon denied the correlation between heat and motion. Then Isaac Newton developed a theory of energy conservation that revealed energy cannot be destroyed or created, as it only ever changes from one form to another, and Einstein's $E = mc^2$ became a commonly known meme that assists general knowledge of energy and matter as essentially different manifestations of the same thing.[8] General memories of energy engineering evolutions inform contemporary appreciations of machinic energetic infrastructure and progress, as they intersect with the biophysical and mental work of everyday energetic realism and capitalist realism. In the age of the Anthropocene, new materialisms posited by Jane Bennett understand energetic entelechy as a life force entwined with the political agency of all things, from rubbish to streams, so that all can be seen as contributing biological life force and soul with communicative potential beyond the limits of human mentality and beyond nature. Current ideas about material consciousness speculate on the existence of "perceptronium," a state of matter in which consciousness exists via variable states of environmental mathematics. Further, a new sixth state of matter called "excitonium," speculated upon for five decades, has been confirmed using energy-loss spectroscopy to reveal quantum processes within a superconductor superfluid state of matter, and it is unknown if this substance is a conductor or insulator of energy.[9] Physicist Roger Penrose and anesthetist Stuart Hammeroff speculate on the existence of quantum microtubules within the brain as a possible explanation for consciousness, energy of mind, and other as-yet-unknown quantum bioneuropowers. The superpower of "telepathy"—a word that has become folksy and unscientific since the decline of psychical research—in the human brain is related to the Visual Word Form Area (VWFA) of the hominid brain associated with mutual gaze, empathy, the mind reading of self-consciousness and consciousness of other minds called Theory of Mind (ToM), or prediction, and precognition. Science not only fails to prove the existence of telepathy, and finds more acceptable words to use, like prediction, ToM, and empathy, but it also fails to define energy and matter once and for all, and art is useful within the Extro-Science Fiction (XSF) of Quentin Meillassoux where science fails but reality continues.

New and speculative materialisms promote more attentive encounters between people-materialities and thing-materialities within an ecological agenda that challenges capitalist materialism. Jane Bennett's *Vibrant Matter* looks at the impersonal affect of human and nonhuman "thingly power" as a nonspiritual and nontraditional vitalism. Bennett's ecological strategy deliberately creates a "capacity for naiveté," such as superstition, animism, and anthropomorphism, and investigates discredited, premodern philosophies of nature.[10] In her theory of agency and assemblage, Bennett turns to the Chinese concept of *shi*, which means style, energy, *élan*, and the potential of things, or, more precisely, the specific arrangement of things, a vibratory assemblage. *Shi* was originally used in military "world-mind-reading" strategy for the ability to "read then ride the *shi* of a configuration of moods, winds, historical trends, and armaments: *shi* names the dynamic force emanating from a spatio-temporal configuration rather than from any particular element within it."[11] *Shi* as *élan*, energy, and potential is also much like entelechy. Bennett engages Henri Bergson's notion of *élan vital* and early work by biologist Hans Dreisch, which connected emerging twentieth-century theories of entelechy, telepathy, and telekinesis. Also relevant to Bennett's vital materialism is Kant's notion of *Bildsungstrieb*, the "formative drive" that attaches to and enlivens dead matter, for the same reasons—it avoids a mechanistic view of formative power and the notion of a soul floating disembodied. Entelechy and *Bildsungstrieb* are neither mechanical body nor ethereal psychical soul. Entelechy animates, and like *shi* it also arranges, designs, and "composes artistically the bodies of organisms."[12] However, unlike the definition of *shi*, Dreisch defined entelechy not as energy but as that which makes use of energy, a physico-chemical material, so that entelechy is rather the labor force of potential that uses materials. This is the case for Bergson's *élan vital*. For Bergson, *élan vital* is drive without design, something akin to groping.[13] For Driesch, vitality is more about the powers of arrangement and direction, while for Bergson it is more about sparking innovation. Vitality relies on powerful materials and their physico-chemical propensity, and for Dreisch and Bergson it should not be reduced to pure materialism.

The identification of energetic vitalism with experimental science helps avoid spiritualization of the vital and/or energetic agent. William James's mystical and spiritual experiences of telepathy under the influence of nitrous oxide—which is also "laughing gas," an anesthetic, an engine enhancer, and a greenhouse gas that is accelerating climate change to dangerous levels via human agricultural use of ammonia in nitrogen-based fertilizers—have provided pioneering insights into modern work on consciousness and psychical research. Nitrous oxide is three hundred times worse for the environment than carbon dioxide, even though carbon dioxide represents 75 percent of greenhouse gases.[14] One ton of nitrous oxide is equivalent to 298 tons of carbon dioxide, and it takes 110 years for nitrous oxide to be removed from the atmosphere, a process that depletes the ozone layer. Carbon dioxide is also associated with altered states of consciousness, hauntings, and experiences of telepathy, and ghostbuster-style psychotherapists routinely check first for gas leaks and proximity to car parks and motorways.[15] Beyond nitrous oxide and carbon dioxide, many other chemicals can offer telepathic insights, and possibly the most powerful are found in plant-based Ayahuasca rituals, which promote an animistic ethos of empathic interspecies and interracial cohabitation by triggering a sense of telepathic connectedness in the mind. However, the brain has psychotropic neuroplastic potential all the time, beyond events of chemical poisoning, intoxication, and medication, and our sense of reality relies on a sense of stabilized chemical balance, along with internal and external environmental effects on mental state that are not particularly dramatic. Energy was also central to James's work on ethics. James's notion of ethics explored the organization of energy, habits, and aesthetics of the self, and it is also aligned to Bergson's notions of *élan vital* and vitalism.[16] My energetic conception of telepathy is anchored in the ancient yet new materialist notion of entelechy and Bennett's notion of a "fabulously vital materiality,"[17] and in the materialist-consciousness problem that has informed the difficult questions of consciousness for philosophy in general.

Bruce Clark, in a chapter called "Energy as Capital," looks at Einstein's discovery of hidden wealth within atoms and all matter; Marx's notion of "theological niceties" of the commodity, which can metamorphize into capital anything with transformative powers, such as energy, vitalism, life, and soul; and the sun as provider for fossil fuel and our modern energy crisis, which cannot be separated from financial crisis.[18] Clark also describes energy as capitalism, and thus as providing for phantasmagoria, occult, and supernatural fantasies and "cosmic capital," but he skips the rare and spooky moment when Marx's analysis of the commodity returns to the problem of brain and spirit. This moment of the return of the brain and spirit in Marx is the crucial connector of commodified energy conversion potential to Hegelian spirit remaining in materialist dialectics, capital, commodity, and the spectral market table.[19] Marx identifies a wooden brain within the market table that is mediumistic and possessed with spiritual table-tapping forces, and it is significant that brain, spirit, consciousness, mentality, and psychology were the exact elements of Hegelian dialectics Marx was so keen to rework. The American oil industry and spiritualism developed at the same time and in close geographic proximity to the "Wild West" of oil communities in Pennsylvania. In America, the spiritualist Abraham James developed psychometric divination to visualize the unseen, deceased loved ones—and oil.[20] Rochelle Rainier Zuck describes Abraham James's spiritualism as a laborer: "He proved infinitely reproducible as an agent of 'practical spiritualism' and was discussed alongside the other drillers, operators, laborers, teamsters, and investors at work in the oil region."[21] In today's oil industry, the Telepath™ has been trademarked as a bespoke analytical tool that can surveil, locate, and repair breaks in subsea oil pipelines or "umbilicals."[22] Oil is the unconscious of modernity, full of magical potential for transformation and wealth creation, as noted in *After Oil*.[23]

Conversely, Sigmund Freud referred to the science of pure telepathy and parapsychology as "the black tide of mud—of occultism." Telepathy can be like black mud and oil because "psychoanalysis has an earth," says Jacques Derrida in his work on "geotrauma."[24] Derrida's "geopsychoanalysis" acknowledges nonanthropomorphic transference situations. Telepathy was at the center of Freudian transference and psychoanalysis, primarily between an analyst and analysand, but it was literally buried by the marketing of Freud's English editor Ernst Jones. Conscious of this history, Derrida in his concept of "geopsychoanalysis" renders the telepathy problem as extended beyond anthropomorphic individual psyches to include nonanthropomorphic geographical boundaries and nation-states because Freud's continental psychoanalytic work on telepathy and the occult were played down by Jones in order to gain "coin and currency" with the "Far West" of Western anglophone intellectual worlds.[25] Geotrauma and occult petroleum wars combining Freud, Marx, Islam, and Big Oil feature in Reza Negarastani's anticapitalist work *Cyclonopaedia*.[26] Earthly, telluric commodities such as oil and gold are almost instantaneously transferable as immaterial and entirely speculative commodities or futures speculation commodities. Geographers such as Steve Pile now take seriously the impact of human immaterial labor, such as futures trading, speculation, and telepathy, and readily incorporate psychogeographic discourse.[27]

Within materialism, scientific concepts of causality overlap with labor-oriented views of energetic cause and effect, as McKenzie Wark emphasizes: energy is the capacity to do work, and it is not *in* coal or oil, but results from labor actions on these materials.[28] Labor has a cosmic dimension when it is understood that work can be exhausted, from an earthly environment or the universe. When energy has been totally distributed, then the laws of thermodynamics refer to this as maximum entropy or "heat death." Both Wark and Catherine Malabou deconstruct dialectical materialism as a problem facing eco- (ecological and economic) consciousness in the age of the Anthropocene: Wark by looking at the Russian revolutionary Alexander Bogdanov (Lenin's rival), and Malabou by deconstructing traditional Marxist conceptions of history and consciousness, and looking at the deep history provoked by the Anthropocene and contemporary neuroscience. In his reading of Bogdanov, Wark suggests energy is a work and we do not know it, just as Malabou says the brain is work and we do not know it, and this phrase might also be applied to telepathy: telepathy is indeed a work of psychic labor and we do not know it. Compared to Lenin (not to mention Stalin), Bogdanov had greater appreciation of the necessity for proletarian learning about biospheric energy systems and Marx's concept of finite natural resources and "metabolic rift," as well as peak oil and climate change problems. Wark points to Bogdanov as an avatar for confronting the Carbon Liberation Front. The Carbon Liberation Front is a direct result of work with fossil fuels but also with Hegelian, Marxist, and capitalist dialectics of nature and history and intellectual progress toward freedom, which in turn provokes a new necessity for social justice for the environment when too much carbon, nitrous oxide, and methane is "liberated." Significantly, Bogdanov also has greater appreciation than Lenin and Stalin of knowledge as organized sensation, and emotional and spiritual dimensions of utopic collectivizations of labor.

Malabou's take on materialist consciousness deploys contemporary neuroscience, which serves to temper the Hegelian spirit remaining within materialist consciousness with her attention to the psychotropics of the technological and neurobiological sublime. Malabou's work on consciousness resonates with Autonomist Vitalist Marxists, such as Antonio Negri, who equate the brain with spirit: "*Geist* is the brain."[29] This is also true for Maurizio Lazzarato and Tizianna Terranova, for whom the immaterial labor of the brain involves action at a distance or telekinetic "noopowers" of immaterial labor, networked society, and societies of control. Malabou agrees that the difference between the brain and psychism is that it "is shrinking considerably, and we do not know it."[30] The neuronal liberation of Malabou's psychical research has a Marxist and neurobiological "spirit," and her work is not spiritualist—her use of the term "psychism" accompanies her concept of "plasticity," which is the human ability to shape the brain-body that shapes the Earth. With rising climate temperatures, the human cyborg brain-body will struggle to cool itself and its essential organs. Inherently sensitive and expensive brain tissue cannot function if its temperature rises by four degrees.[31] Malabou views the problem of human consciousness and the Anthropocene within deep geological history and concludes that humans have become structurally alien due to their brain work, and that what is now required is a new form of consciousness that is in fact a *lack of consciousness* and a reorientation of our habits and addictions to the Anthropocene in order to adapt.[32] Cognitive architecture for the Anthropocene may proliferate shiny futurist spaceship utopias, fascist dystopias stemming from Steve Bannon's Biosphere 2 strategies, and spherical geodesic EcoDomes that put a silvery space-age bandage over extraction pollution while multiplying extraction and product efficiency. What was once science fiction is now becoming a reality, and as novelist Sharman Apt Russell speculates, the Anthropocene may accelerate pantheistic consciousness of telepathy.[33] Any additional predictive, preemptive, and telepathic module or novel cognitive trait evolving in response to cosmic heat death entropy and chaotic climate of the Anthropocene, as well as the pressures of increased automation and artificial telepathy, will require both a metabolic price and fuel. For our human cyborg animal, which has become increasingly alien to Earth, the development of new cognitive abilities—and the maintenance of existing everyday brain functions—will still require humble sugars, and, significantly, "the new trait must enable our animal to find this extra amount of energy in its environment."[34]

Malabou reflects on consciousness itself as a geological reality in which humans have become a geological and telluric force, and she reminds us that lakes and mountains are not foreign to the ecosystem of the human brain, and if humans destroy them, they destroy their brains and themselves as well. Further, the neuroplastic brain is addicted to the environment through chemicals and synapses, and thus brain and culture together must be seen as mind altering and psychotropic.[35] Malabou recognizes that "psychotropic" is a strong word, and she anchors this term in discussions of human biochemical addiction to techno-scientific power and the deep evolutionary history and neuroplastic mind-altering habits of human neurophysiology.[36] Neuroplasticity is the human brain's ability to modulate habits that "shape the environment that in turn shapes our brains."[37] Neuroplasticity is in a sense increasingly a world and geological power or superpower, with almost telekinetic and telepathic agency to shape the environment that shapes the brain. The human brain is extended technologically, as described by Marshall McLuhan, to encompass the planet, and we are now addicted to technology. Now in the Anthropocene the human brain's role as a social organ is accelerated as it works with immaterial labor *telekinetically* and collaboratively with other organic brains and inorganic extensions and materials, as described by Maurizio Lazzarato and Tizianna Terranova.[38] The collective brains' immaterial psychopower/neuropower works at a distance as an extended multitude of brains, things, and technology that act in the Anthropocene like glaciers and snow to modulate the behavior of humans and the materiality and energy of the Earth.

Telepathic Art for Neuroplasticity and Solarization

Figure 13.1 shows four video stills of environmental performance tests undertaken with Marissa Benedict, David Reuter, Heather Ackroyd, and Dan Harvey in 2016, while engaged in the BRiC "On Energy" residency, and Naomi Oliver and myself in 2017 in full-body alfoil (aluminum foil), while engaged in the EDACC (Energy, Data Abstraction, and Cognitive Capitalism) residency. All participants performed with an alfoil hat on their heads, an artistic prop borrowed from my participation in Swiss Italian artist Gianni Motti's 2016 international telepathic celebration of the birth of the art movement called Dada. I wished to further explore the alfoil hat as it relates to conspiracy theories of mind control and as a trope associated with climate science denial. All participants in my video experiments are clearly *against* climate science denial, so the performance is absurdist and ironic. I also wanted to give attention to the site of skull and brain and to develop conceptual resonance with the shiny, silvery built environment of the rural industrial landscape and especially the aluminum EcoDome of the LaFarge Exshaw cement works. For me, the space-age materiality of alfoil is nurturing, like a silvery emergency blanket used to modulate temperature and protect food, energy, and medication. Alfoil acts as a conductor of electricity, and the alfoil hats facilitate a connection to the invisible cognitive architecture of the EcoDome. My work here with processes of collaborative coupling, twinning, and group work involves and/or suggests mental transference—psycho-bio-energetic phenomena or telepathy—as a form of "energy dialogue."[39]

I am also interested in how humans may be communing with the nonhuman environment, in particular with the EcoDome. My work with the EcoDome feeds into my exploration of Buckminster Fuller's energetic dome aesthetics of cognition, radio and electrical waves, and telepathy, as Fuller discusses in his introductory text to Gene Youngblood's book *Expanded Cinema*. My collaborative alfoil performance with Oliver was undertaken in an area of the Blue Mountains National Park that was once used for coal and oil shale mining, and in our work we created a noise performance in response to the rich sound ecosystem of birdlife in that area. We were almost entirely wrapped in alfoil "spacesuits" and stood upon an emergency blanket. We also resembled the Tin Man from *The Wizard of Oz* who stiffens in the rain while cutting down trees and can only be animated by oil. I imagine these fragile astronauts in tattered alfoil to be communing with Wark's "Commie aliens confronting peak energy and climate change problems" in Alexander Bogdanov's *Red Star* science fiction, as well as the telepath characters of *Solaris* and Sharman Apt Russell's

Fig. 13.1

Anthropocene science fiction. The bottom artwork in this figure shows my work with Diderot's map of the human circulation system from Vaclav Smil's chapter on human energetics[40] overlaid with aluminum foil disks that echo the metal disks of the stentrode, as well as an atomic fall of atoms or the clinamen of shiny metallic EcoDome-factory-spaceships. Atomic matter is central to understanding the hard problems of energetic reality and materialist consciousness, as atomic spin intertwines material and immaterial particles and forces. The silvery disks are also like the silvery solar cells, and this connection points to my interest in solarization within photography and conceptual art fictions, as well as in mental and industrial processes. My work reflects on the modulation of sunlight as essential for bioneurological health as well as social, industrial, ecological, and economic health.

Figure 13.2 captures the inside of the EcoDome (top) and a series of video stills sequenced from an edit of video documenting the outside and inside of the EcoDome. The supply chains of power cables, roads, and rail transport facilitating the extraction of rock and limestone also feature. I explored the EcoDome as a hypnotic, dazzling, and spectacular presence within the landscape, alternatively nicknamed "Thunderdome Spaceship" by some locals. I learned of the role the dome plays in minimizing the dust pollution of the Lafarge Exshaw cement works, and the extremely high-energy-intensive process of converting rock into cement—rock is heated to three times the heat of the sun, and the facility requires its own electrical grid.

Figure 13.3 in the top two images shows the EcoDome juxtaposed with my handcrafted UFO architecture installed at the Cementa_13 art festival, a festival designed to revitalize a small town that has lost the biggest cement factory in the Southern Hemisphere. As in other works selected here, I have made links between landscapes used for coal and limestone extraction through work in both Australia and Canada via my engagement with the Blue Mountains National Park area of New South Wales and Banff National Park and Rocky Mountains in Alberta. The second two images show a graph of overseas oil petroleum imports into Australia and then my upside-down appropriation of these graph data into an oil painting titled *Oil on Oil* (2017). *Oil on Oil* is a datascape of accelerated information interwoven with televisual blur, and made of acrylic and oil paint on board. Brushstrokes reveal the coded infoscape through the prism of Australian oil imports from Congo, Nigeria, Algeria, United Arab Emirates, New Zealand, "Other Countries," Malaysia, Thailand, Indonesia, Philippines, Papua New Guinea, Brunei, "Asian Group peak 12/2007," "Peaking group 11/2005," and "refinery closures." The third row of images document a work called *Energy Séance* staged on the Bow River, and it

Fig. 13.2

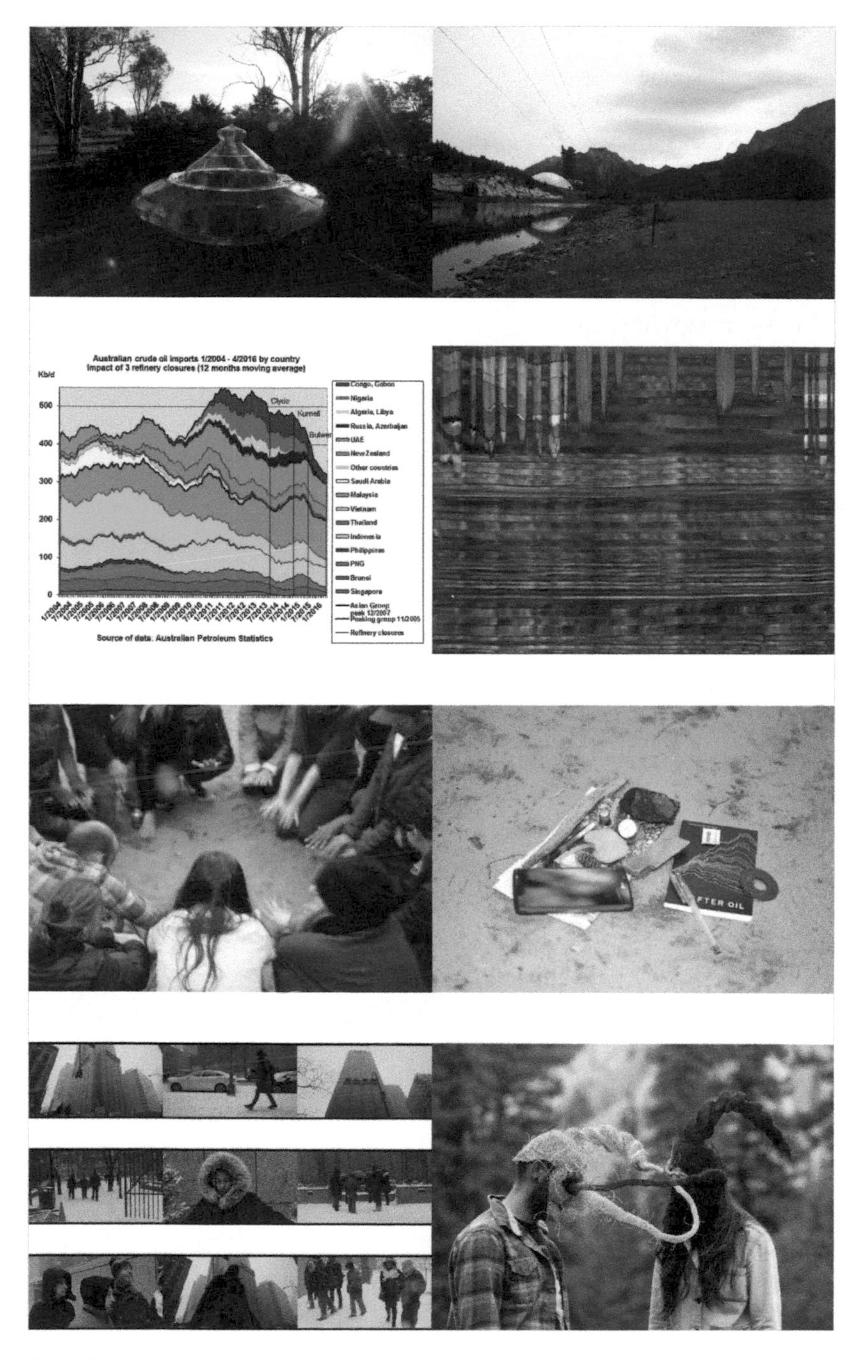
Australian crude oil imports 1/2004 - 4/2016 by country
Impact of 3 refinery closures (12 months moving average)
Kb/d
500
400
300
200
100
0
Clyde
Kurnell
Bulwer
Congo, Gabon
Nigeria
Algeria, Libya
Russia, Azerbaijan
UAE
New Zealand
Other countries
Saudi Arabia
Malaysia
Vietnam
Thailand
Indonesia
Philippines
PNG
Brunei
Singapore
Asian Group peak 12/2007
Peaking group 11/2005
Refinery closures
Source of data. Australian Petroleum Statistics
FTER OIL

Fig. 13.3

captures a moment of collectivized bio-energy of group circle work to suggest energy transference through centralized gaze and outstretched fingers simulating the pose of psychic mediums. Many participants contributed "energy objects" placed in the center of our circle, including a mobile phone, rocks, coal, coins, a plastic pen, and the book *After Oil*. The fourth line of images shows my recent work with data, telecommunications, energy, and séance performance. *Data Centre Séance* (2017) and *Energy Séance* (2016) were conceptualized about the same time. *Data Centre Séance* involved a séance at a data center called the Longlines/Titanpointe Building in Manhattan. The snow blizzard accentuates the bio-energetic vulnerability of the participants as they communed with each other and with this ominous windowless data center with large vents releasing the enormous heat energy generated by data storage. For *Solstice Telepathy* (2016), I wove telecommunications wire through a coiled warp and weft. Any coiling of copper wire creates a very primitive radio set of extremely low frequencies (ELF). ELF is a form of radiation at the low-energy end of the electromagnetic spectrum that some scientists speculate may be an earthly explanation for some paranormal events. *Solstice Telepathy* continues my experimentation with the energetic and ELF-telepathic connection between people. The artwork is about binding two people within a conjoined black-and-white headpiece. Performance of the weaving occurred on the evening of a northern summer solstice, and this cosmic event informed the title *Solstice Telepathy* and my understanding of communion between the skull-sphere and the Earth-sphere as a metaphor for extended and materially engaged cognition that is both human and nonhuman. The yin and yang of hemispheric solstice light and dark extremes mapped onto heads abbreviates a wide spectrum of mental energies. Data storage and telecommunications, as well as cement factories, are well-known huge consumers of energy and emitters of greenhouse gases, and my work with woven telecommunications wire and data centers, as well as psychogeographic relations between people, objects, and architecture, is in part connected to my attempt to rewire and solarize the extended social brain away from fossil fuel dependency.

Figure 13.4 shows my work with black-and-white solarized photographs documenting the rural industrial landscape of the EcoDome and related Lafarge cement industry facilities, as visible in the top image. The second row of images depict a waterside factory emitting pollution into the atmosphere, as well as an atmospheric pollution monitoring station at the foreground of the water, with the factory visible in the distance. Canada is ahead of Australia in terms of transitioning away from coal power by mostly using hydropower, yet coal still contributes to electrical power in Canada. Electricity and fossil fuels are

Fig. 13.4

represented through inversion of light and dark within a black-and-white-only photographic spectrum, with gray midtones pushed toward black (using Adobe Photoshop). The definition of solarization is inhibition of photosynthesis, and photosynthesis involves metabolic reactions and reduction of carbon dioxide using energy absorbed by chlorophyll from sunlight. Solarization of these images acknowledges metabolic rift and the Carbon Liberation Fronts of Bogdanov/Marx/Wark. Etymologically the term "solarize" also means to convert to solar energy instead of fossil fuel power, and my photos wish for further reduction of fossil fuel use in general. I also hope my solarized images may in a sense be projected onto society, and help find a mental and actual industrial switch from fossil fuels to solar and renewable power. There is no liberation for Earth from the dominance of the Sun and its powers, only liberation of the brain from fossil fuels to stop the release of hydrocarbons that trap heat and cause climate change. Solarization is not infrared or thermal imaging, but it is uncanny, nonretinal coding of light information suggestive of a technological imaging of energy waves and other forms of "invisible" radiation. Solarization almost evokes a spiritualized transcendental transformation of rays of light into a mysterious dark smog materiality, and the blacks suggest the omnipresence of fossil fuels. The telepathic Anthropocene message of dead plants and animal presence in fossil fuels is surely that humans are accelerating climate change and species extinction. The final, bottom image of this plate depicts the Lafarge Exshaw car park and electrical power grid. The cars and trucks are parked in emergency mode for fastest possible exit, as the massive energy requirements of the facility require constant vigilant attention to the possibility of a major catastrophic explosion. The power grid facilitating the extremely high energy demands of cement production is the structure at the center of the car park, with the noses of vehicles parked facing away, ready to flee an energy disaster.

The top image in figure 13.5 is an installation detail from my *Unconscious Extraction* exhibition (2017) where my black-and-white conjoined headpieces are combined with two red and gray weavings of head and brain structures hovering above a pile of brown-black coal taken from Clarence Colliery next to the Blue Mountains National Park. The top red weaving is a headpiece with rounded skull and facial features of eye sockets, protruding nose and lips, mouth orifice, jaw, and cheeks. Wires extending from the head suggest artificially animated veins and energetic unpruned excess of synapses. Just below this woven head is a machinic and encephalitic brain with split left and right hemispheres and a red spinal column. These works self-consciously engage an artistic dialogue with the neural interface of the stentrode brain sensor that accesses the brain via cardiovascular conduits. I use a contradictory "low and slow" technology

Fig. 13.5

of hand weaving to render high-tech entrepreneurial neurocapitalism as plastic and within the hands of delinked artists. Also in the back of my mind is the science fiction of Bogdanov, who pioneered blood transfusion, developed Marx's concept of metabolic rift and early industrial climate consciousness, and challenged Lenin's counter-revolution that resulted in Stalinist dialectical materialism. Bogdanov's early science fiction story about climate change, *Red Star*, imagined "subtle and invisible" threads connecting delicate brains and indestructible machines.[41] The multiple images below situate a performance of artistic neuroplasticity and energy dialogue between two artists, Colin Wintz and Annie MacMillan. As I discuss in my paper "Neuromodulations of Extro-Scientific Telepathy," psychic energy, empathy, and the visual word formation part of the brain, which is also the part of the brain that is predictive and preemptive, are activated by mutual gaze. My woven headpiece enables humans to playfully perform their struggle to maintain eye contact and empathic connection through the umbilical mesh of telecommunications that extend the brain. In the case of Oxley's stentrode, wires are beginning to be inserted literally within the brain via neural mesh, and the stentrode safely enters the brain through veins as a tiny, expandable lattice of wires and tiny, round nitinol (nickel-titanium) sensor disks.

As an artist, I identify with Keller Easterling's extrastatecraft strategy when I bring neuroplasticity to my work with problems such as energy, telepathy, and materialist consciousness. Easterling identifies the need for artists, designers, and thinkers to bring a "change module" or switch within the social brain to trigger alternative design contagions, and, like me, she appreciates social theorist Gabriel Tarde's important point on how social networks are run by fictions of suggestion and crowd contagion. Tarde's special attention to crowd contagion is also his work on the telepathy of crowds. Easterling works with a push-pull engagement with the mid-century cybernetics and the ecology of mind thought of Gregory Bateson.[42] Easterling is against the gridlock of modernity, yet sees precisely in gridlock the occasion to game or hack the cybernetic system in order to look for opportunities to change the multiplier or means of production and create new contagions.[43] Crucial to Easterling's call to activate urgent change are the strategies of doubles, multiples, imitation, and mimesis as informed by Tarde, and Easterling's strategy informs my own work with doubles and mimesis. Malabou, Wark, and Easterling assist my artistic thinking on the binding of physical and neurological structures to climate, energy cultures, and cybernetic systems. Like artists working with the aesthetics of Fuller and Bateson, these theorists point to invisible architecture[44] as a new vision of climate itself as a form of architecture.

My work on energy, telepathy, cognition, and materialism is grounded in a battle with corporate and government control of energy-materialism and mind-body. As an artist I can use the independent plasticity of the plastic arts to imagine energy together with telepathy and materialist consciousness (and all their crossovers and contradictions) as forces within my thinking and thinging, ideas and art, and thus give new power to imagination and neuro-/psycho-plastic subjectivity in response to energy and fossil fuel problems. Future investigations stemming from my work here may include a work on *The Telepathy of Marx*, in which I combine Malabou's ecosophy with the Hegelian brain-spirit haunting of Jacques Derrida's *Specters of Marx*, within new experiments in art, writing, transhumanist collaboration, and organizing.

Notes

1. Stuart Hameroff, "Consciousness, Neurobiology, and Quantum Consciousness," *Quantum Mind*, http://quantum-mind.co.uk/theories/penrose-and-hameroff/consciousness-neurobiology-and-quantum-mechanics; and "Dr. Stuart Hameroff on Quantum Consciousness and Moving Singularity Goal Posts," Skeptico, http://www.skeptiko.com/stuart-hameroff-on-quantum-consciousness-and-singularity/.
2. I have identified artists Frantisek Kupka, Marina Abramovic, and Robert Barry, and theorists Sigmund Freud, Wilhelm Reich, and Rosalind Krauss, as working with telepathy as a form of energy transfer. "Transference" is a mental, electrical, and material science term, and this is highly significant for Barry, Reich, and Krauss. See my articles "Human and Non-Human Telepathic Collaborations: The 60s, 80s, and Now," *Colloquy: Text, Theory, Critique* 22 (2011): 139–64; and "Neuromodulations of Extro-Scientific Telepathy," in *Psychopathologies of Cognitive Capitalism: Part Three*, edited by Warren Neidich (Berlin: Archive Books, 2017), 319–65.
3. I explore "telepathology," which is a medical term for the diagnosis of blood disease at a distance via digital technologies, in relation to telepathy as it interfaces with the hard sciences of biology and physics, and I coin this term "telepathologies of cognitive capitalism" in my article "All That Is Solid: Speculative, Quantum and Cognitive Aesthetics of Telepathy and Telekinesis," *Leonardo Electronic Almanac* 22, no. 1 (2017): 91, 95, 96, 97–99.
4. For a recent update of my PhD thesis on telepathy in art (available online), see the latest on Artbrain "Art and Telepathy," curated by Jacquelene Drinkall and Warren Neidich, Artbrain, http://www.artbrain.org/.
5. Bronac Ferran, "Neuromorphobia (Hypehypehyper)," in *Psychopathologies of Cognitive Capitalism: Part Three*, edited by Warren Neidich (Berlin: Archive Books, 2017), 116.
6. Cited in Ferran, "Neuromorphobia," 120.
7. Vaclav Smil, *Energy in Nature and Society: General Energetics of Complex Systems* (Cambridge, MA: MIT Press, 2008), 2–3.

8. Barry Lord, *Art and Energy: How Culture Changes* (Arlington, VA: American Alliance of Museums Press, 2014), 2.
9. Siv Schwink, "Physicists Excited by a New Form of Matter, Excitonium," 7 December 2017, Physics Illinois, https://physics.illinois.edu/news/article/24114.
10. Jane Bennett, *Vibrant Matter: A Political Ecology of Things* (Durham, NC: Duke University Press, 2010), 18.
11. Bennett, *Vibrant Matter*, 35. Like Deleuze's use of the word "adsorption," it brings together elements while retaining the integrity of each element.
12. Bennett, *Vibrant Matter*, 71, 78.
13. Bennett, *Vibrant Matter*, 79.
14. Peter Grace and Louise Barton, "Meet N_2O, the Greenhouse Gas 300 Times Worse than CO_2," *The Conversation*, 9 December 2015, https://theconversation.com/meet-n2o-the-greenhouse-gas-300-times-worse-than-co2-35204.
15. Carrie Poppie, "A Scientific Approach to the Paranormal," *TEDxVienna*, 3 March 2017, https://www.ted.com/talks/carrie_poppy_a_scientific_approach_to_the_paranormal.
16. Sergio Franzese, *The Ethics of Energy: William James's Moral Philosophy in Focus* (Frankfurt: Ontos, 2008).
17. Bennett, *Vibrant Matter*, 63.
18. Bruce Clark, *Energy Forms: Allegory and Science in the Era of Classical Thermodynamics* (Ann Arbor: University of Michigan Press, 2001), 43–47.
19. Karl Marx, *Capital, Vol. 1, Part 1: Commodities and Money*, http://www.marxists.org/archive/marx/works/1867-c1/ch01.htm#S4. See also my catalogue text *Eco-Spirit*, available on https://unsw.academia.edu/JacqueleneDrinkall.
20. Rochelle Raineri Zuck, "The Wizard of Oz: Abraham James, the Harmonic Wells, and the Psychometric History of the Oil Industry," *Journal of American Studies* 46, no. 2 (2012): 313–36.
21. Zuck, "The Wizard of Oz," 314.
22. Terry Stebbings et al., "Leak Location in Subsea Umbilicals," conference paper, "OnePetro 2007," https://www.onepetro.org/conference-paper/SPE-108956-MS.
23. Petrocultures Research Group, *After Oil* (Morgantown: West Virginia University Press, 2016).
24. Jacques Derrida, "Geopsychoanalysis: ' . . . and the rest of the world,'" *American Imago* 48, no. 2 (Summer 1991): 200.
25. Derrida, "Geopsychoanalysis," 200–201.
26. Jacquelene Drinkall, "Anthropocene Aesthetics of Telepathy and Action-at-a-Distance in New Materialisms," conference presentation, "XHums: Aesthetics after Finitude," University of New South Wales, 5–6 February 2015, http://www.niea.unsw.edu.au/sites/default/files/AAF%20Conference%20Program.pdf.
27. Steve Pile, "Distant Feelings: Telepathy and the Problem of Affect Transfer over Distance," *Transactions of the Institute of British Geographers, Royal Geographical Society* (with the Institute of British Geographers) 37, no. 1 (2011): 44–59.
28. McKenzie Wark, *Molecular Red: Theory for the Anthropocene* (London: Verso, 2015), 25.
29. Charles Wolfe, "Cultured Brains and the Production of Subjectivity: The Politics of Affect(s) as an Unfinished Project," in *The Psychopathologies of Cognitive Capitalism, Part Two*, edited by Warren Neidich (Berlin: Archive Books, 2014), 253.

30. Catherine Malabou, *What Should We Do with Our Brain?* (New York: Fordham University Press, 2008), 8.
31. Sanford Kwinter, "Neuroecology: Notes Toward a Synthesis," in *Psychopathologies of Cognitive Capitalism, Part Two*, edited by Warren Neidich (Berlin: Archive Books, 2014), 321.
32. Catherine Malabou, "The Brain of History, Or, The Mentality of the Anthropocene," 17 February 2017, www.youtube.com/watch?v=wJPLGEdRGGc.
33. Sharman Apt Russell, *Knocking on Heaven's Door* (New York: Yucca Publishing/ Skyhorse Publishing, 2016).
34. Thomas Metzinger, *The Ego Tunnel: The Science of the Mind and the Myth of the Self* (New York: Basic Books, 2010), 43. Thanks to Warren Neidich for this valuable reference.
35. Catherine Malabou, "The Brain of History, Or, The Mentality of the Anthropocene," *South Atlantic Quarterly* 116, no. 1 (2017): 45. Special thanks to Christopher Malcolm and Jeff Diamanti for references, and for references to the telepathy of the art collective Ant Farm.
36. Malabou, "The Mentality of the Anthropocene," 46–49.
37. Bruce Wexler, *Brain and Culture: Neurobiology, Ideology, and Social Change* (Cambridge, MA: MIT Press, 2006). See also discussions of Wexler in Neidich's *Psychopathologies of Cognitive Capitalism.*
38. Maurizio Lazzarato, "The Concepts of Life and the Living in the Societies of Control," in *Deleuze and the Social*, edited by Martin Fugslang and Bent Meier Sorensen (Edinburgh: Edinburgh University Press, 2006), 180; Tiziana Terranova, "Another Life: Social Cooperation and A-Organic," *Digithum* 12 (2009), http://digithum.uoc .edu/ojs/index.php/digithum/article/view/n12-terranova/n12-terranova-eng.
39. Jacquelene Drinkall, "Human and Non-Human Telepathic Collaborations," *Colloquy: Text, Theory, Critique* 22 (2011): 152–53; Charles Green, "Group Soul: Who Owns the Artist Fusion?" *Third Text* 18 (2004): 180; Sigmund Freud, "The Uncanny" (1919), in *The Standard Edition of the Complete Works of Sigmund Freud*, edited and translated by James Strachey, vol. 17 (London: Hogarth Press, 1955), 234. Green and I highlight this sentence from Freud on energetic psychic transference: "This relation is accentuated by mental processes leaping from one of these characters to another—by what we should call telepathy—so that one possesses knowledge, feeling and experience in common with the other."
40. Smil, *Energy in Nature and Society*, 119–146.
41. Wark, *Molecular Red*, 10; Alexander Bogdanov, *Red Star: The First Bolshevik Utopia*, translated by Charles Rougle, edited by Loren Graham and Richard Strites (Bloomington: Indiana University Press, 1984), 65.
42. Gregory Bateson greatly informs Buckminster Fuller and artist architects working with energy and telepathy, such as Juan Downey and Ant Farm.
43. Keller Easterling, *Extrastatecraft: The Power of Infrastructure Space* (London: Verso, 2014); author notes from 2016 BRiC "On Energy" lecture, workshop, and individual consultation with Easterling.
44. I do not have space to detail the invisible architecture of Juan Downey or his text *Architecture, Video and Telepathy: A Communications Utopia* (1977), but I can recommend this text: Julieta Gonzalez, "Juan Downey's Communications Utopia," http:// curatorsintl.org/images/assets/julieta_gonzalez.pdf.

PART III

The Politics of Energy Culture

CHAPTER 14

Rejecting Solar Capitalism

Jenni Matchett

Energy from fossil fuels is central to how global culture functions materially and economically. By almost every definition, this culture is unsustainable. Despite some acknowledgment by the private-sector energy industry of the way this culture "reinforces the systems of exploitation, dispossession, and domination already dismantling the possibility of a future for the majority of the planet's inhabitants," we have yet to figure a viable path forward that feasibly addresses the coming reality of an uninhabitable planet.[1] The pace and magnitude of system change are compromised by our (in)capacity to balance the tensions among historic precedents, current "needs," and visionary future outcomes. Energy transformation exemplifies this impasse. Beyond dispute is our need for a zero-carbon world, and yet our only way of knowing how to exist in the world stands in direct contradiction to meeting that goal.[2] In light of this contradiction, and within a society that has determined solar energy to be one of the few readily available technologies capable of delivering drastic emissions reduction, we must attempt to reconcile our emissions reduction goals with radically social objectives if we are to be successful in transitioning our energy system and in creating a culture of conservation that isn't at odds with itself. But how? There is still time to critically evaluate our current renewable energy deployment practices and attempt to abandon the parts of these practices that perpetuate the negative cultural externalities of energy capitalism. Solar energy gives us the opportunity to design a renewable energy paradigm that transforms the cultural values made impossible in a fossil fuel economy. I believe not only that this will accelerate our energy transition, but that a sustainable transition requires it.

To support this objective, I discuss three deployment models that have emerged in recent history. These models help frame solar energy circumstances in the contemporary moment. They shed light on the ways in which solar deployment in its current forms perpetuates petroculture and, more dangerously, validates it. In short, they are part of the practical foundation we must renovate if we are going to generate a new paradigm in relation to energy use.[3]

Everything discussed here uses the North American electricity grid as landscape. The complexities of the electricity system are plentiful and regionally unique, and the broad North American electricity system discussed here is no exception. However, it's a space where the impacts of solar energy deployment are tangible and is therefore relevant to the questions of energy transition considered below.

Electricity

To understand and evaluate the promise of socially driven solar deployment in energy system design, we must understand the infrastructure system that solar energy deployment relies on—the electrical grid—as well as the legislative and economic parameters within which solar electricity comes into existence.

The electricity grid was designed more than a half-century ago and continues to function in a fashion true to its original design. That design is characterized by large, centralized electricity generation sources and vast networks of high-voltage transmission lines that transport electricity from its generation point to its point of consumption. Electrification accelerated economic expansion and is, of course, vital to modern comforts. However, what makes the electrical grid remarkable (and a contender for the most important invention of the twentieth century) is that there has been very little change in its design despite decades of technological transformation and developments in alternative energy. A vast assortment of regulatory, technical, transmission, and distribution players coordinate themselves to maintain this archaic infrastructure, which delivers reliable electricity and/or fuel supply and services to wherever there is demand.[4] Recently, however, decayed grid infrastructure and pressure from a carbon-constrained world have forced electrical grid operators to embark on a transition process.[5] Important to this transition, and in stark contrast to how electricity generation has been deployed in the past, is how pro–solar policy directives have favored the individual electricity consumer (that is, the household), generally in the form of financial subsidy. In this way, incentivized consumer energy choice is now an important facilitator in the deployment of solar energy.

Depending on location, renewable electricity development is dictated by experimental legislative action and the creation of market mechanisms in the form of subsidies or other financial incentives. This imperfect policy experimentation and financial market dependency limit our ability to imagine a solar-powered culture, specifically because of how these two elements adhere to economic norms. The deeply entrenched rules of finance that govern our energy economy mean the deployment of solar must conform to the financing norms of historic energy infrastructure deployment. Subsidies and incentives are written in a language that large institutional capital understands. As the monetary value creators in our global economy, those who control capital are the ones who determine the rules by which renewable assets generate value (and who therefore receive the most economic benefit).[6] Further, solar electrons generally get on the grid starting with a legislative process that sets the amount of renewable energy supply that a distribution utility must procure to meet electricity demand.[7] Because of this, electrical distribution utilities are the gatekeepers of electricity system transformation, and interaction with these monopolistic bodies is an inevitable reality of contemporary solar electricity deployment. Putting it simply, an arbitrary determination (in relation to targets that will actually equal zero carbon) of regional emissions reduction targets, along with solar incentives that generate explicit economic value for capital providers, form the two bases on which solar energy is currently deployed.

Electricity is an obscure cultural form. Its invisible and ubiquitous nature creates this obscurity, making it difficult to discern what is and what is not a cultural product of the electricity system. Consideration of the ways in which the electricity system historically and technically overlaps with the free-market financial system gives insight into why we have not yet been able to create a renewable energy paradigm. The overlap between these systems has generalized a culture of consumption and fossil energy dependence, both of which have been vital to the increasing dominance of the free-market financial system. Electricity enables the construction of identity through consumption.[8] What, how, where, and why we consume are all increasingly dependent on electronic devices, a fact that is highly visible in our current cultural ecosystem. Electricity has, therefore, become a primary tool in the way we curate our consumer-driven lives. Electricity provides us with the basic hardware required to consume, and it also determines the conditions for consumption. Recognizing that the electricity sources that power our current consumption ecosystem pose a significant threat to life as we know it, the dominant approach and instinct has been to replace those sources with non–carbon-dioxide-emitting alternatives rather than to supplant the cultural legacy, habits, and contingencies of the grid. Understanding the electrical envelope this way helps explain why we can't imagine solar outside of the discourse of carbon reduction, and why renewable energy is not yet a paradigm in and of itself.

Within the existing paradigm of electricity production and consumption, several solar deployment models have emerged, including solar energy as consumer product, as asset for retail investment, and as energy access equalizer. I will discuss each of these in turn.

Solar Energy as Consumer Product (Tesla)

Tesla has brought branded solar into mainstream consumer culture. Though the sales model relies on the material consumption habits of the individual, the company has created consumer brand allegiance to an electricity generation source—a new development in energy culture.

With overwhelming shareholder support, the premier electric vehicle manufacturer has created the first fully integrated consumer energy company by merging its original electric vehicle and energy storage business with its rooftop solar–installing "cousin," SolarCity (run by cousins of Tesla's founder, Elon Musk). This is an important moment in the evolution of personal energy choice. Currently, the companies that provide consumers with the option to make personal solar energy choices fall into three categories: (1) the utility; (2) large private business (nationwide rooftop installers, such as SunRun, Sungevity [Sungevity filed for bankruptcy in March 2017], or Vivint.Solar); and (3) small private business (local rooftop installers). At present, consumer brand recognition for solar energy is relatively low.

Geographic variation in energy policy and pace of policy change add another layer of complexity to the solar provider's quest to connect with the consumer market. As discussed, solar deployment is determined by regional and local policy, which means that the choices available to consumers regarding where their power comes from are not uniform. Further, deployment is dependent on the distribution utility, a circumstance that limits an energy provider's flexibility to deploy consumer engagement tactics. Not all energy consumers may even want the option to choose solar, but if they do, the differentiation in market regulations can hinder consumer awareness and product scalability.

Though policy and utility changes are required for the successful deployment of solar, those changes tend to be relatively slow. We have not yet seen what a consumer energy revolution looks like when a solar brand becomes a household name. Enter Tesla. In a relatively short period of time (compared, for example, with the time the Ford Motor Company and other automotive behemoths have had to capture consumer brand allegiance), the brand is gaining significant consumer traction. The SolarCity brand may go unnoticed, but for those who have consumed any current news in the last year, most have almost assuredly heard of Tesla and have definitely heard of Elon Musk.

But now there is no more SolarCity. There's just Tesla. And its rooftop solar products will be branded Tesla, too. Which means solar energy has just gained a household brand name.

Solar Energy Asset as Retail Investment (Wunder Capital)

The Wunder Capital platform provides individual investors with the option to finance solar project development through a retail investment. Though this model preserves capital inequities inherent to all sectors of the investment industry (in this case, unequal power over critical infrastructure deployment and unequal access to the economic benefits of solar energy in the form of interest), it's also an important departure from traditional energy investment, in that it allows individual energy consumers to directly influence renewable development with their investment dollars.

Until very recently, retail investment options in solar energy were few. Formerly, the options were either to produce the substantial sum of cash required to purchase a solar system for one's own dwelling or business, or else to make a stock-equity investment in a publicly traded solar company, like Sun Edison (largest solar developer in the world until 2016, when the company filed for bankruptcy) or SolarCity (acquired by Tesla in 2016). In comparison to their fossil fuel counterparts, solar companies have suffered for several reasons, not least of which is their inability to economically perform in a culture of short-term financial metrics . . . the irony.

Because of this, the odds are still stacked in favor of fossil fuel deployment from the perspective of retail investment options. Registered Retirement Savings Plans (RRSPs) and 401Ks hold mutual funds stocked with the world's largest energy-development companies, most of which generate the majority of their revenue from fossil fuel–related activities. These companies are financially stable, have proven longevity, and provide reliable returns. In other words, they remain the obvious choice for the average retail investor.

Wunder Capital falls into a new category of retail investment class in which investors can finance the development of solar energy assets and earn returns based on long-term, stable project revenues. Wunder does all the back-end work to get a solar asset development deal done. It sources the project, negotiates the investment terms with the solar developer, facilitates the retail investment in the project, and manages those returns, which will, of course, vary jurisdictionally. This investment model emerged in a landscape where investors are demanding the option to participate in solar project finance, and where traditional capital sources are unable to serve this high-growth market sufficiently.

It is important to note that these investors are not directly receiving the electrons generated by their investment; instead, they earn a financial return on the monetary value the asset generates.

Solar Energy as Energy Access Equalizer (Community Solar)

Community Solar, in theory, provides everyone in a community with the option to choose solar generation over other sources, regardless of their economic or residential circumstances. Though utility system constraints at present prohibit Community Solar in its current form from becoming a more prevalent phenomenon, the model has inspired a new wave of energy equality advocacy and consumer energy education—organizations that support solar energy politics being defined by the Left.

Community Solar is most commonly defined by regulatory statute as a solar energy system that allows multiple energy consumers to enter into a long-term agreement with a solar asset owner to receive financial benefit (usually in the form of electricity savings) from the project. Alternatively, it allows multiple energy consumers to actually purchase a small portion of a local solar facility and receive the economic and environmental benefits from that portion of the facility. The former is a more common deployment model, due to the legal complexity of allowing multiple individuals to own distinct portions of a utility-scale solar array.

This model was created with the intention to provide consumers access to solar, regardless of their socioeconomic status. In theory, Community Solar allows any individual living near the solar project to become a member of that project via a third-party energy provider. In most instances, by signing up for Community Solar, one is entering into an agreement with the solar asset owner that currently resembles the long-term (generally twenty years) power purchase agreement model originally created to facilitate the deployment of large-scale renewable energy at the distribution level.

Community Solar member benefits are dictated by the solar incentives present in that region. Under current rules, this means members receive the same credit on their utility bill for their Community Solar share size as they would receive for a similarly sized system on their rooftop (the idea is that the share covers the majority of the consumer's energy needs). The value the customer actually receives is equal to the difference between what the solar asset owner charges for the solar credit that the utility must allocate to the customer's account, based on his or her share of the solar system's generation, and the credit the customer receives from the utility. The price the solar asset owner charges is generally less than the credit the customer receives, so the customer is guaranteed savings. This means individuals can access the benefits of solar energy as defined by the regional policy in which the Community Solar project is being deployed.

In practice, the model struggles in a system that is designed to discourage models of a collective energy commons. Very few jurisdictions have actually implemented the policies required to develop this collective model, and even in those places where Community Solar deployment has been successful, consumer requirements, utility reliance, and skewed incentives mean that the "power" often sits far from the individuals who make up a community. In this way, we see hints of similarity between this innovative approach to democratizing energy and the cooperative rural utility model of 1930s America that promised (and failed) to deliver the social transformations that it made possible.[9] Still, the Community Solar model hints at an equitable and renewable energy path forward.

Consumer product. Retail investment. Community Solar: models that highlight "a situation where everything changes and at the same time nothing changes."[10] Very little has changed, culturally, if solar energy deployment is most powerfully influenced by material consumption (Tesla), unequal capital returns (Wunder Capital), or potentially problematic approaches to collective energy (Community Solar). If there exists a hope to shape the energy transition through which we are currently living—to both reduce emissions and support the creation of culture that is "collective, equitable and just in all of its practices and principles"—solar energy is that hope, because it is the most promising and ripe energy space to critique and inform.[11] As a technology that has quickly surpassed predictions in both cost decline and pace of deployment, the solar revolution will prevail. Whether or not anything changes depends on our ability to confront solar capitalism.

Notes

1. Not an Alternative, "Institutional Liberation," e-flux, November 2016, http://www.e-flux.com/journal/77/76215/institutional-liberation/.

2. Petrocultures Research Group, *After Oil* (Morgantown: West Virginia University Press, 2016), 17.
3. Keller Easterling, *Extrastatecraft* (New York: Verso, 2014), 232.
4. United States Department of Energy, *Grid Modernization Multi-Year Program Plan*, 2015, xiii, 4, https://energy.gov/sites/prod/files/2016/01/f28/Grid%20Modernization%20Multi-Year%20Program%20Plan.pdf.
5. Christopher Knittel and Ignacio Pérez-Arriaga, *Utility of the Future* (Cambridge, MA: Massachusetts Institute of Technology, 2016), 137–39, http://energy.mit.edu/wp-content/uploads/2016/12/Utility-of-the-Future-Full-Report.pdf.
6. MIT Energy Initiative, *Future of Solar* (Cambridge, MA: Massachusetts Institute of Technology, 2015), 114, http://energy.mit.edu/wp-content/uploads/2015/05/MITEI-The-Future-of-Solar-Energy.pdf.
7. Galen Barbose, *U.S. Renewables Portfolio Standards Annual Report* (Berkeley, CA: Lawrence Berkeley National Laboratory, April 2016), https://emp.lbl.gov/sites/all/files/lbnl-1005057.pdf.
8. David E. Nye, *Electrifying America: Social Meanings of a New Technology, 1880–1940* (Cambridge, MA: MIT Press, 1990).
9. Abby Spinak, *Infrastructure and Agency: Rural Electric Cooperatives and the Fight for Economic Democracy in the United States* (Cambridge, MA: MIT Press, 2014), 40.
10. Daniel Barber, "Architecture," in *Fueling Culture: 101 Words for Energy and Environment*, edited by Imre Szeman, Jennifer Wenzel, and Patricia Yaeger (New York: Fordham University Press, 2017), 49.
11. Imre Szeman and Jeff Diamanti, "Strategies for a Left Energy Transition," *Canadian Dimension* (February 2017): 56–58, https://canadiandimension.com/articles/view/beyond-petroculture-strategies-for-a-left-energy-transition.

CHAPTER 15

The Switch

Keller Easterling

In the comedy of errors that is US car culture, contradictions and cross-purposes seem to thrive. The mid-twentieth-century monovalent highway network replaced a finely grained rail network to conform to default modernist scripts claiming that newer is better. In perennial cycles of obsolescence and replacement, new infrastructure technologies have overwritten existing networks, however intelligent or efficient they may be. Adhering to false logics of traffic engineering that linked roadway width to traffic volumes, those highways were doomed to both continually inflate in size and never relieve congestion. While reliance on fuel provokes military entanglements and destroys the planet's atmosphere, car manufacturing has also been treated as a key indicator of employment and economic health. And driving—often a dreary and time-consuming chore performed by gripping a wheel and staring forward at a road—is, at least in car commercials, stubbornly portrayed as a luxury—the sexy freedom of rounding curves on wet roads.

As President Donald Trump's Eisenhower-era highway plans and environmental deregulation meet the newest shiny technologies associated with "smart" autonomous vehicles (AVs), what will be the next engineering paradox or myth of luxury? While Trump points to a future in the past, the same modernist narratives still claim that newer is smarter, even if the organizations they create are effectively dumber. Celebrating another false logic, AVs are projected to create a "mobility internet"—an internet of things—in which synchronized digital devices attached to everything will finally make the stiff spatial world fluid.[1] Again, replacing incumbent networks, new technologies will provide the right answer.

AV technology is heralded as transportation's magic bullet, but it is unclear how the advent of the technology will be organized. In ideal projections of a fluid, smart, digital world, AVs in efficient platoons will perfect driving by increasing mobility, saving fuel, reducing emissions, and increasing productivity.[2] But if all the major car companies develop automated technologies and sell AVs in the same way they sell the individually owned family car, congestion and vehicle miles traveled (VMT) will only increase. Organizing the AVs in fleets, at first, seems to drastically decrease the number of cars needed. And with no need to pay for the labor of drivers, AV fleets would also initially appear to reduce the cost of travel.[3] But there is a boomerang effect. Fleets of ridesharing AVs might reduce vehicle miles traveled. However, because of increased convenience and reduced cost, passengers might take AVs instead of transit, so the number of cars on the road could skyrocket. This is just one of several simple spatial variables to consider. Expand the size of every person in the commuter train or subway to the size of a car, and it is clear that the congestion would even overwhelm the counter-effects of carpooling and platooning. Larger cities are already experiencing increased congestion because riders are choosing Uber or other ridesharing companies over transit. Consider another spatial variable: if driverless cars increase productivity by allowing passengers to work while traveling, and if there is then less incentive to shorten the commute, AVs might also encourage sprawl. When the magic bullet ricochets, the optimistic predictions turn to dire warnings of increased congestion, emissions, and sprawl.[4]

How, then, do the automobile industry, the sharing economy, and transit organize their efforts in the face of such uncertainties? And how does the United States fund repairs for the broken transportation infrastructure on which all of these efforts rely?

While the habitual response may be to look for a "solution" in the next emergent technology, like futuristic flying cars, an alternative response might be to *alter a relationship* or to *rewire the network* with a *spatial variable*—an architectural volume that acts like a switch when placed between existing transportation technologies.[5] Whatever the promise of digital technologies, space is itself a technology, an information system in play, as well as an underexploited medium of innovation. However heavy and lumpy, it does not need digital devices to make it dance. Digital and spatial networks can make each other smarter, but they can also make each other dumber. The *switch*, an intermodal node for upshifting and downshifting into transportation of different capacities, may offer the best model for organizing maintenance, innovation, and investment in this shifting transportation ecology.

Imagine the switch located at a transit stop in the suburbs. That stop might have a low ridership because potential riders cannot drive their car to the stop and leave it in a parking lot because the car is needed for other purposes throughout the day, or because the train doesn't take riders to their final destination on the other end of the journey. Faced with these options, a commuter might simply take a car from door to door. But when linked by a switch, AV fleets and transit become reagents with a chemistry that redoubles efficiencies. AVs can deliver a vastly increased ridership to transit, and serve any number of trips that would require a household to own multiple vehicles at rush hours, while also handling the last mile of travel.

In the morning, a suburban commuter could walk or take an AV to the nearest switch, upshift to transit, arrive in town, and walk or take another AV to the office. Other family members could organize their trips to work or school in a similar way. And on either end of the trip, any of these riders could arrive at a switch and downshift instead of upshift. The commuter might take a bicycle from an urban station to the office or, on the way home, change clothes at the switch and jog the last mile home.

Linking AVs and transit at the switch reduces congestion, sprawl, and emissions; makes transit healthier; increases AV efficiencies; and potentially even consolidates liabilities. There have been concerns that vehicles in fleets might often be traveling empty, but at the switch, if they can both deliver and pick up, they will rarely be without passengers. While there are many imponderables about liability when individually owned cars are driverless, with switches and vehicle fleets, liability issues at least resemble familiar forms like those for trains or, as some have suggested, elevators.[6]

The switch is also an urban space and a cultural institution that, by changing habits, has ramifying effects on the shape of cities and suburbs. Having to switch during a journey is usually treated as an inconvenience, but consider a space like Grand Central Station in New York City. More than a train station, it allows for shifting between networks and modes of transportation, and its internal programs (food markets, restaurants, bars, banks, gyms, and other retail) respond to time-sensitive needs at different moments of the day. Similarly, the switch could become the hub of complicated household itineraries and activities—the place for food shopping, exercise, violin and ballet lessons, day care, and monitored spaces for students to gather after class and practice or study. Each activity provides foot traffic for the other. Even if AVs could be perfectly synchronized to function with optimized efficiency, the distances between dispersed destinations and errands would increase vehicle miles traveled. Consolidating some of the primary repetitive errands to do with school, work, child care, grocery shopping, and households adjusts a spatial network to make both spatial and technological networks smarter.

By concentrating businesses and errands, the switch can also be a real estate organ capable of providing revenues to augment the now uncertain public funds for transportation. With real estate revenues from its train stations, Japan Rail, for instance, has been able to impeccably maintain and upgrade its system while also funding research and development in technologies like Maglev. Transit organizations are sitting on large amounts of land that is only used for parking, but AVs require much less space. Transit stations make some money from parking, but if that space were converted to a switch and filled with retail and business opportunities, it could return much larger revenues to support transportation innovations of all sorts.

The switch might also address issues to do with employment and cultural divides caused by poverty or racism. Transit systems are typically routed to serve more affluent neighborhoods, meaning that travel times for those who live in poorer areas are too long on a good day and impossible on a day with other difficulties. But AVs can travel anywhere, and the switch can be a place for the cultural mixtures that generate diversity as well as social and economic cohesion. Even now, the routines of the switch can be rehearsed with the vehicle fleets of the ridesharing economy with the view to better transitioning employment once cars are fully automated. The switch may be a place to organize that transition, since its concentration of businesses and services is a source of potential jobs to compensate for those lost in the taxi and ridesharing businesses.

The switch is not another utopian proposal. It is an idea that could be gamed and exploited in ways that would destroy it, and in the United States cars can perhaps only be pried from the cold, dead hands of their owners. The advent of AVs could simply lead to the next transportation catastrophe, or to mobility companies organized as monopolies with bulletproof accumulations of power and information. But maybe the idea of the switch simply represents a shift in habit of mind or the essential disposition of power in cities—a sense that there is more information, choice, and equity for more people when there is interplay between digital, spatial, and human networks.

The switch might then encourage a different, impure approach to political activism. An activist can righteously fight against attitudes about energy and mobility in the United States by protesting environmental deregulation, fighting for a retightening of things like corporate average fuel economy (CAFE) standards, or working to establish alternative renewable energy sources. Or the switch could be a social prescription of well-meaning urbanists bent on "transit-oriented development" (if the association with markets and real estate was not already too tainted for the purist). But these sorts of approaches have little chance of outwitting the PR of the current political climate mitigating our transportation stupidities.

When thinking about not only a single new technology but the mixing chamber for many technologies, a designer can design volumes with shapes and outlines as well as the spatial and cultural medium in which they are suspended. Since there is hardly any hope of making a system shift like the switch without also being able to manage the spin that surrounds it, this *medium design* works on many networks simultaneously. It shifts potential in a matrix with instructions and specifications for a spatial architectural proposal, and it also crafts the promotional stories that attend the idea. Those stories are not an end in themselves, but rather a deliberate instrument to manipulate habits and routines in a way that has measurable spatial consequences.

And since the messages of car commercials have managed to create a contagion that has utterly changed the urban landscape with a desire, maybe a contagious story about the switch might travel under the radar and galvanize developments around alternative political and environmental alignments. It would be a good trick if designers somehow directed the next wave of cloying, soft-focus commercials for mobility companies, rather than cars. While having to stop and change modes of transportation would ordinarily be regarded as a disruption to the dream of seamless travel, these ads would promote another dream in which the switch is portrayed as the new marker of luxury, elegance, and freedom.

In thirty seconds, an advertisement might show a single-parent household with a two-year-old and babysitter at home and a teenager in school. On a bad day, when the babysitter has to go home early, she takes an AV to the switch, waits for the teenager's soccer team to be delivered, and places baby brother in big brother's care, before running to catch a train. With baby brother on his shoulders, big brother waits for Mom's train while winning approving looks from the teenage girls at the switch. When a relieved Mom joyously arrives, all three pass through the glorious, voluminous spaces of the switch. There are other parents in the background picking up from day care and escorting children dressed in karate uniforms or carrying violin cases. Given that AVs provide access for many neighborhoods, including those that might previously have been without mobility, the crowd in the switch is diverse. Mom and her two children shop for dinner in the market, and then get into an AV in the queue that is picking up departures. In the background, others are climbing onto bicycles or beginning a jog for the last mile home. Maybe, in the distant background of the shot, there is even a city that no longer has parking spaces. On their way home, teenage brother plugs in his smartphone and blasts his music while a super-cool, hands-free Mom, now facing her children instead of the road, juggles oranges from the grocery bag.

Another soft-focus ad might feature the switch as a new micro-institution in suburban neighborhoods. In places where there are no trains, or in any neighborhood that wants to aggregate riders for AV pooling, there might be a small switch within walking, running, or biking distance to many surrounding homes. It has a monitored waiting station for travel to and from school, and it is the place where everyone gets their AV to go to work or to a larger switch that has a train. It has bike storage and changing rooms, but it might also have day care, food stalls, farmer's markets, exercise classes, or work spaces. This ad would show Dad returning from work to the switch. While picking up his child from day care, he hears someone call to him. Mom has just arrived and signals that she is going to change into her running clothes. Dad hurriedly buys dinner and gets to the bike racks. The next bit of advertising cuteness shows Dad cycling home with the child in a front carriage while a jogging Mom challenges them to a race. Merging the sexiness of ads for running shoes with ads for mobility, exertion has become the new luxury.

The wholesomeness continues. The landscape the family is moving through has a very different morphology. Because there is a nearby switch, there is really no need for the 40-to-60-foot-wide roadway onto which the suburban house usually fronts. Nor is there a need for garages. Just as the car was a contagion that inflated distances in the twentieth-century suburb, now the AV contracts those distances and replaces them with greater density or more collective landscapes like parks or—if the ad is especially trendy—the newest forms of suburban agriculture.

However impure, promoting stories about switching or interplay might be a sly form of activism, one with the capacity to have compounding effects on energy, the environment, mobility, and settlement patterns. With political superbugs, whether they are car companies, oil conglomerates, or presidents, earnest proposals can be ineffective, and direct opposition sometimes only delivers the nourishing rancor that fuels a fight. But the switch satisfies the desire for market incentives before these powers have a chance to politicize them, and the superbugs themselves are abundant proof of the power of stories to bend steel and buckle concrete. If it is better to keep the power players guessing, the switch might be a cunning way of shifting the ground at a precarious political juncture.

Notes

"Switch" was originally published by e-flux Architecture as a part of *Positions* on September 8, 2017.

1. Earth Institute Columbia University, "Transforming Personal Mobility," 27 January 2013, http://sustainablemobility.ei.columbia.edu/files/2012/12/Transforming

-Personal-Mobility-Jan-27–20132.pdf. See note 1 in that document. The term "mobility internet" was coined by William J. Mitchell, Christopher Borroni-Bird, and Lawrence D. Burns, *Reinventing the Automobile: Personal Urban Mobility for the 21st Century* (Cambridge, MA: MIT Press, 2010).

2. Matthew Claudel and Carlo Ratti, "Full Speed Ahead: How the Driverless Car Could Transform Cities," McKinsey & Company, August 2015, http://www.mckinsey .com/businessfunctions/.
3. Earth Institute Columbia University, "Transforming Personal Mobility."
4. *Autonomous Vehicle Technology: A Guide for Policy Makers* (RAND, 2016). See also http://www.rand.org/pubs/research_reports/RR443–2.html.
5. Keller Easterling, *Organization Space: Landscapes, Highways and Houses in America* (Cambridge, MA: MIT Press, 1999).
6. "Self Driving versus Driverless—A Mobility Update with Alain Kornhauser #CES2107," interview, https://www.youtube.com/watch?v=_HucrtH1h4U.

CHAPTER 16

Beyond Carbon Democracy: Energy, Infrastructure, and Sabotage

Darin Barney

In the opening line of his essay, "Virtuosity and Revolution: The Political Theory of Exodus," Paolo Virno writes, "Nothing appears so enigmatic today as the question of what it means to act."[1] Virno was thinking of advanced capitalism in general; his observation arguably applies doubly to petrocultures and their impasses, in which most conventional forms of political action seem blocked, gestural, empty, complicit, or weak in the face of the unretractable character of petrocapitalism (even beautiful Norway tripled production in the face of the truth of climate change). Climate change—petrocapitalism's evil twin—bears similarly on the prospects of action. If global warming is what Timothy Morton has described as a "hyperobject," a phenomenon so "massively distributed in time and space relative to humans" as to appear unscalable, then what is left for the action of human subjects besides the hypocrisy and humiliation to which Morton refers?[2] "In order to break the spell," Virno goes on to say, "we need to elaborate a model of action that will enable action to draw nourishment precisely from what is today creating its blockage."[3] In this chapter, I will explore whether sabotage points toward such a model in the context of our contemporary energy culture.

I am indebted to Timothy Mitchell and his book *Carbon Democracy* for drawing my attention to sabotage as a way to think about politics in the context of fossil fuel economies and polities. The basic claim is straightforward, expressed succinctly by Mitchell:

> Political possibilities were opened up or narrowed down by different ways of organizing the flow and concentration of energy, and these possibilities were enhanced or limited by arrangements of people, finance,

> expertise and violence that were assembled in relationship to the distribution and control of energy . . . [It] was the movement of concentrated stores of carbon energy that provided the means for assembling effective democratic claims.[4]

Mitchell's story concerns coal, the era of progressive democratic and welfarist reform in the West, and the transition to an oil-based energy and economic system. In it, the material properties of coal and its infrastructures left the movement of coal (that is to say, its value as a *commodity*) vulnerable to saboteurial disruption by organized workers, who succeeded in leveraging this disruptive power to secure progressive democratic concessions from capital and capitalist states. In the late nineteenth and early twentieth centuries, organized workers involved at various points in the transportation of coal in Europe and North America used sabotage to extract a broad range of progressive, structural concessions from industrial capital and their governments. These actions were typically led by militant miners' unions, but often spread to, or were coordinated with, railway workers, dock workers, and marine workers positioned to disrupt the flow of coal through foot-dragging, slowdowns, tampering, and strikes. According to Mitchell, their power "derived not just from the organizations they formed, the ideas they began to share or the political alliances they built, but from the extraordinary quantities of carbon energy that could be used to assemble political agency, by employing the ability to slow, disrupt, or cut off its supply."[5]

Among the benefits that workers in North America and Europe secured by leveraging their ability to sabotage the movement of coal, Mitchell lists the extension of suffrage; rights to form labor unions, to strike, and to create workers' political organizations; and labor reforms including the eight-hour day, social and unemployment insurance programs, protection against job loss due to accident or sickness, and the first public pension schemes. As Mitchell puts it, "working people in the industrialized West acquired a power that would have seemed impossible before the late nineteenth century."[6] In response, capital turned to oil, a commodity whose material properties lent themselves to movement via infrastructures (pipelines and tankers) that required less human labor and were more flexible, thereby reducing the impact of disruption and undermining or sabotaging the growing political power of organized workers.

Mitchell's account figures the scene of carbon politics as a scene of sabotage. At the close of his book, he makes specific reference to the current state of extreme or unconventional petroleum extraction. He writes: "To transform kerosene-impregnated rock formations and bitumen-filled sands into oilfields is to acknowledge that what we call nature is a machinated, artificial territory in which all kinds of novel claims and political agencies can form."[7] The question this prompts us to ask is whether it is productive to think of these "novel claims and political agencies" in terms of sabotage.

Typically, when we think of sabotage, we think of someone breaking or physically destroying something, which is what allows for the reduction of sabotage to violence, criminality, and even terrorism. However, sabotage is only contingently related to the specific tactics by which it is carried out in any given circumstance. Sabotage has a very long and diverse history, in which the tactics and techniques deployed by saboteurial actors range widely.[8] Sometimes this has involved breaking things, but often it has not. As Arturo Giovannitti wrote in his jail-cell introduction to Émile Pouget's classic 1912 text *Sabotage*, "It is not destructive. It has nothing to do with violence, neither to life nor to property. It is nothing more or less than the *chloroforming of the organism of production*."[9] Elizabeth Gurley Flynn of the Industrial Workers of the World (IWW), in her crucial 1916 text on sabotage, defines it as "the conscious withdrawal of the workers' industrial efficiency."[10] Even the capitalist state has recognized that sabotage cannot be defined solely in terms of violence. In a landmark 1927 ruling, the US Supreme Court upheld the conviction of an IWW organizer found guilty of sabotage merely for possessing pamphlets that advocated "going slow" (a classic tactic of nonviolent workers' sabotage) on the grounds that "any deliberate attempt to reduce profits in the manner [of slowing down on the job] would constitute sabotage."[11] In the case of going slow, sabotage disrupts value that is still potential and not yet actual by reducing the amount of value transferred by labor power to the commodity in the process of production. Even these erstwhile antagonists would seem to agree on what defines sabotage: it is a form of action that intervenes in established structures and systems of distributing and accumulating value, especially processes of extraction, production, reproduction, and circulation. Sabotage is a form of action that withdraws, disrupts, or subtracts from the efficiency of work and the efficiency of flows. Thus, it can take many forms, including many that have nothing to do with violence or the destruction of property.

In what follows, I will highlight three things that Mitchell's account helps us to specify about sabotage in relation to a possible politics of or through energy in the context of petroculture and climate change. I will then consider a few attributes of sabotage that exceed the account given by Mitchell.

Sabotage Is Normal

It is a long-standing conviction in the intellectual and political history of sabotage that capitalists are its primary and most committed practitioners. In making her case for workers' sabotage, Flynn refers to the case of Frederic Sumner Boyd, who was arrested during the Paterson Silk Strike of 1913 for suggesting that textile workers should enter the dye houses and adulterate the silk with chemicals to render it unweavable. Flynn points out that Boyd was merely advocating "something that is being practiced in every dye house in the city of Paterson already, but is being practiced for the employer and not for the worker."[12] As William Trautmann put it in another IWW pamphlet from 1912, "sabotage is in daily use for the enlargement of capitalist profit interests."[13] Similarly, in his 1921 treatise, *The Engineers and the Price System*, Thorstein Veblen (upon whose discussion of sabotage Mitchell relies heavily) demonstrates that sabotage by businessmen is part of the "ordinary conduct of business" in "any community that is organized on the price system." Profitable business under market conditions thus demands and comprises a "voluminous running administration of sabotage."[14]

Mitchell invokes this theme in order to bolster his argument that, whether via the manipulation of price through the throttling of supply or the development of pipelines to undermine workers' power to disrupt flows, petrocapitalists are themselves agents of sabotage. This observation extends easily into the present era and beyond the energy sector to include the planned obsolescence that is a structural condition of the digital economy and the evasion of environmental standards that keeps the wheels of automobility turning.[15] If sabotage is a structural feature of the organization of power in a capitalist economy and society—whereby the flow of value is manipulated in order to maintain or elevate price or to secure the strategic advantage of firms—then it is fair to say that this form of sabotage is a normalized attribute of capitalist economies, though it is typically styled as innovation or competitiveness. As suggested above, the association of sabotage with criminality (as opposed to enterprise or political action) is strictly social, not definitive or necessary.

The Mode of Sabotage Is Immobility (Not Publicity)

Sabotage figures the political sphere as a scene of movement or mobility, not a sphere of appearance and discursive interaction. Sabotage thus strikes at the heart of the commodity (which must move to generate value) and is a primarily nondiscursive mode of action particularly suited to a context in which the norms and practices of publicity have been largely drained of integrity.[16] From this perspective, if the public sphere is a sphere of communication, it is a sphere of communication-as-transportation. Under these conditions, the site of political intervention shifts from the transmission of meaning to the circulation of bodies and things.[17] It is telling, in this regard, that the history of sabotage by organized workers in the West begins not in a factory but, as Flynn points out, on the docks. As Evan Calder Williams emphasizes, the site of sabotage's earliest political potential was not manufacturing but, instead, "the dockyards, the train lines, all the juncture sites of circulation."[18]

This, too, is clear in Mitchell's account of sabotage and the movement of coal: sabotage is a mode of action strategically addressed to mobility—in this case the mobility of commodities—rather than to publicity, even in cases where it instrumentally resorts to publicist tactics from time to time. This suggests, for example, that the strategic point of Canadian antipipeline demonstrations or legal actions is not to persuade people with arguments, but to make it impossible to move oil sands bitumen.[19] In Mitchell's account, states were coerced by forced immobility into conceding to workers' demands; they were not *persuaded* by the force of better arguments. In this respect, we might say that the political mode of sabotage is not dialogical but, instead, logistical.[20]

The Medium of Sabotage Is Infrastructure

Sabotage is a form of action that finds its medium in infrastructures of circulation, transportation, and mobility. As Mitchell shows, coal cars, petroleum pipelines, and oil tankers can be more or less vulnerable to sabotage, depending on their particular material configuration and attributes. This implies a conception of media, infrastructure, and even communication that exceeds what prevails when politics is reduced to communicative action in the form of intersubjective dialogue, and when the critical questions concern how media "distort" democratic communication and whether new infrastructures or media might support more authentically democratic modes of communication.[21] From the perspective of sabotage, the significance of infrastructure as the medium of political action is not the quality of symbolic exchange or dialogue it supports but, rather, the opportunities it presents or denies for the disruption of material flows.

In these three respects—that sabotage is a normal mode of capitalist value accumulation, particularly in the energy sector; that sabotage is a politics strategically oriented to mobility, especially the mobility of commodities; and that infrastructures for this movement provide sabotage with its distinctive media—Mitchell's account helps us to think about the character of many of the "agencies and claims" at play in contemporary struggles over energy development and transition. However, I am less sure that with Mitchell we arrive at the more radical heart of what sabotage might signal in terms of political action and subjectivity in a carbon economy and polity. Up to this point, the politics characterized as saboteurial actually sound quite conventional: identify a vulnerable choke-point in a commodity-chain; organize and appear in force to block it; make a demand; keep choking until that demand is met with either concession or violence. Do it again. This is the story of carbon democracy that Mitchell tells so well.

If that is all there is to sabotage, it names many of the forms of action we see in efforts to disrupt expansion of the petro-economy, including physical blockades and occupations of testing sites and pipeline routes, disruptive intervention in public hearings and regulatory proceedings, assertions of territorial sovereignty and custodianship, litigation to contest approvals and licenses, and publicist tactics aimed at depriving extractive enterprises of social license (or at least driving up its cost). These are all effective forms of political action, and they are saboteurial in many ways. And, as with the transitions to coal and petroleum, they might even succeed in prompting capital to accelerate its transition to new energy sources and infrastructures that will sabotage the growing power of the "petrotariat."[22] For all that, it is still not clear that these are especially novel forms of action or that they embody what is really unusual and radical about sabotage as a particular form of action and subjectivity. Mitchell's account tells us something about sabotage, but not everything. Sabotage might be a much stranger and more complex form than what has been discussed so far in this chapter, portending a much more radical form of political action and subjectivity, one less easily identified and located in the present scene of petropolitics.

With this in mind, I will highlight three other attributes of sabotage, informed by an account that has been emerging in the work of Evan Calder Williams, arguably the most interesting and thoughtful contemporary scholar of sabotage.

Sabotage Is Internal to Capitalism

It is not just that sabotage is practiced by capitalists against the interests of competitors, workers, and consumers, and is therefore a normal structural attribute of capitalism. It is that when sabotage happens, it comes from inside the system it attacks, not from without. The potential of sabotage is already present in any interdependent system. Because sabotage is already internal to the system of capitalist power, it is readily available for deployment against that system provided one is ready to bear the considerable risks of enacting it. As Williams writes, sabotage is "the deployment of a technique or activation of a capacity, at odds with the apparatus, system or order within which it is situated and for which it was developed."[23] Sabotage is potentially effective as a mode of contention because it is already there inside the system being contested.

Quoting Flynn, Williams observes that all that is required to activate sabotage against, rather than in favor of, the existing order of things is

> the "fine thread of deviation": the impossibly small difference between exceptional failure and business as usual, connected by the fact that the very same properties and tendencies enable either outcome. If we are to think of sabotage as a process that negates productivity, it's a negation that can't be disentangled from the structures of productivity itself.[24]

In this sense, sabotage is "politics as judo" in that its organizing principle is the use of subtle shifts in position to divert an opponent's force against them. It works from the inside out. This is why, as Williams observes (cleverly quoting a 1987 master's thesis by US Air Force captain Howard Douthit): "History does not point to an effective countermeasure to sabotage."[25] The suggestion here is not that counter-hegemonic sabotage is easy or risk free for those who might summon the courage to undertake it. If there is a difference between capitalist sabotage and anticapitalist sabotage, it is that the former is rewarded by markets and mostly protected by law, while the latter is subject to prohibition and punishment. Instead, the point is that in deeply embedded and highly interdependent systems such as petrocapitalism, the resources—the technologies, agencies, and relationships—for potential disruption, resistance, and deviation are already present, even if only latently. In this sense, sabotage stands counter to the cynical position that, under conditions of petrocapitalist totality, meaningful action is impossible prior to a fundamental transformation that is beyond our capacity to effect. By contrast, a saboteurial disposition assumes that systems based on inequality and contradiction already contain energies that harbor the potential to undo them. Releasing these energies by pulling fine threads of deviation is what it means to act.

Sabotage Is Mediation

If infrastructure is the medium of sabotage, then sabotage itself is a practice of mediation. Throughout the history of its thought and practice, it has been well understood that the potential of sabotage relies on an unspoken confederacy between the saboteurial actor and the systems and materials he or she leverages in the course of committing disruptive action. While sabotage takes the form of a withdrawal, it is not simply a strike or a collective withdrawal of the single element of workers' labor from a productive system (indeed, historically, sabotage arose as a tactic under conditions where the simple withdrawal of labor proved ineffective against companies that could simply replace that labor). Saboteurs do more (or less) than remove their labor: they disrupt the generation, circulation, and accumulation of value in a system by acting in it, with it, through it, in concert with any number of its many and complex animate and inanimate elements.

This is the sense in which sabotage is materialist: it rests upon an assemblage of human and nonhuman agencies, a confederacy of workers, sites, architectures, machines, processes, and materials. Williams is particularly insightful on this point:

> Sabotage . . . invokes a very slippery, inhuman solidarity and communication with the very things you fight against . . . In order to sabotage, one must know the landscape, one must know the factory, one must know the home, one must know the hospital, *intimately*. It is not a knowledge that can simply come from afar—a Molotov doesn't count. It's knowing precise points of failure, and so what it means is this odd complicity with anonymous and yet non-neutral materials.[26]

What distinguishes sabotage is an intimate relationship with the complex materiality of the situation—a materiality composed of a multiplicity of human and nonhuman constituents whose assembly is what generates the saboteurial possibility.[27] In Williams's account, sabotage extends from simply going slow to include any act that "use[s] elements of a machine, system, organism, code, network, or city against its designed function" in order to disrupt or defect from capitalism's hold on the organization of production, time, space, and life.[28] Sabotage thus materializes a form of "invisible organization" that inheres in everyday experience and skill, and embodies a relationship of intimacy between human actors and the nonhuman things (materials, processes, technologies) that are complicit in their action. Sabotage, we might say, is essentially relational, and, therefore, it is definitively a practice of mediation.

In this sense, sabotage represents a specifically political form of what Richard Grusin has called "radical mediation." For Grusin, "mediation operates not by neutrally reproducing meaning or information but by actively transforming human and non-human actants, as well as their conceptual and affective states."[29] Radical mediation, he says, "should be understood not as standing between preformed subjects, objects, actants, or entities, but as the process, action, or event that generates or provides the conditions for the emergence of subjects and objects, for the individuation of entities within the world."[30] Sabotage is mediation in the sense of what Williams refers to as a relationship of "odd complicity," even if not all mediation is necessarily saboteurial. Sabotage is mediation doubly radicalized, mediation in its most explicitly political form. From the perspective of radical mediation, the two-part question facing the saboteur is as follows: What are the specific constituents, relations, and orientations that might deviate from, destabilize, and destructure the hegemony of petrocapitalism, and through what practices and processes of mediation might they be assembled and materialized?

Sabotage Is Postsubjective

In this light, we might well ask: who, in this wild scene of material complicity, is the responsible actor, and what is she/he/it doing? He/she/it is implicated in an assemblage, but definitely not standing up *in an assembly* and making a speech, writing a blog post, or casting a vote. Like the capitalist saboteur acting anonymously as "the economy" or "the market" by orchestrating a conspiracy of human and nonhuman elements, and unleashing their effects across unbounded temporal and spatial distances, he/she/it might never appear in public in his/her/its *own right* at all. Sabotage is a way of being political that reverses the republican logic of politics as the event of appearance and the act of speaking.

"Sabotage," according to Williams, "is a constant activity, it is not something that will happen . . . but a way to wage war without ever coming into the open." Sabotage is not evental but everyday—"not an act with a definite social content," Williams writes, "but rather an exacerbated relation" that places the saboteur in a constant state of withdrawal, or being-in-refusal, even as they might appear to be or to say otherwise.[31] This is what distinguishes Williams's account of sabotage from Mitchell's, in which sabotage, though enacted on flows and mediated by infrastructure, is still more or less indistinguishable in its basic logic from the strike. In a strike, disruptive action is collectivized, but its authors are declared, as are the demands whose satisfaction will end the disruption and restore operations to normal. Strikes can be saboteurial, but they are not necessarily or always so.[32] Sabotage is more akin to what Stefano Harney and Fred Moten, referring to the struggle over the means of social reproduction in and by the black undercommons, call "planning": "the plan is to invent the means in a common experiment launched from any kitchen, any back porch, any basement, any hall, any park bench, any improvised party, every night . . . planning in the undercommons is not an activity, not fishing or dancing or teaching or loving, but the ceaseless experiment with the futurial presence of the forms of life that make such activities possible."[33] Here, planning is not the discrete action of a particular individual or collective but the mediated emergence of the conditions of possibility for action at all. Planning does not happen; it is under way. This points to what is most radical about sabotage: its confounding of the regime of the modern, liberal, humanist subject who is seen and said to act only when the boundaries of his or her action can be established and he or she can be named, represented, and held accountable for it.

In reducing political action to public speech, Hannah Arendt famously writes that, "without the disclosure of the agent in the act, action loses specific character . . . action without a name, a 'who' attached to it, is meaningless."[34] Indeed, sabotage is not a form of action fit for the Athenian polis or the attenuated liberal democracies that aspire to its dialogical ideal. Sabotage, according to Williams, is "a form of social war opposed not just to a global order of reproduction, circulation, and management but also to the most basic structures of representational politics that order strongly encourages us to adopt."[35] Sabotage explicitly eschews the diluted forms of political action provided by liberal democratic publicity. "Sabotage," Williams says, "is anathema to any notion of representation, of voting, of being a citizen and above all the notion of the human that underwrites the entire project and therefore structures what people understand the political to mean."[36] This is not simply because the saboteur would sooner write a disabling bug into a game console's code than sit through a steering council meeting or organize an encampment. It is because the structure of sabotage—its "invisible" organization, the temporal and spatial gaps between the act and its effects, the manner in which the act combines multiple human and nonhuman agencies with which the saboteur collaborates but which he or she cannot control—destabilizes the very idea of action as a discrete event that can be attributed directly and openly to an individual actor who is its exclusive author. In Williams's account, sabotage is inherently (and not just tactically) clandestine, a category of action that is defined in opposition to the modern liberal conceit that an act is a bounded occurrence for which an individual can and must "appear" to take responsibility or credit. "Sabotage," he says, "necessarily insists on acts that must not be traced back to their source. They therefore become properties of the world."[37] The difference between action as a property of individuals and action as a property of the world is the difference between the liberal and the saboteurial subject.

Capitalists and other producers of inequality and exploitation have always been saboteurial subjects in exactly this sense: their acts become properties of the world that are not traced back to their source. When a capitalist in a Manhattan office is prompted by an algorithm to close a plant in a small Ontario town because some combination of subsidy, tax relief, and the price of labor is more advantageous elsewhere, he or she acts politically by leveraging systemic interdependencies to withdraw or withhold the firm's contribution to the circulation of value.[38] The capitalist commits a great and disruptive sabotage, but its effects are felt far from where and when he or she "acts," and it is not clear that the person is acting alone or that his or her "decision" can even be understood as a discrete action attributable uniquely to him or her. It implicates a complex assemblage of other human and nonhuman elements, and it is less of an event for him or her than it is a "constant activity," a "property of the world," and so the capitalist never "comes into the open" and the act is never "traced back to its source." He or she is never named, is not responsible, is not held accountable. He or she is a saboteur. The designation of action as a property of individuals mediated by speech is central to liberalism, but it has only ever been applied selectively and has never adequately described how power operates or how effective political action actually happens. This is revealed most dramatically under the conditions of contemporary neoliberalism, wherein some individuals (the poor) are required to stand up and take responsibility for that over which they have no control or influence, while those who have the most power and influence over events (the rich) are relieved of any responsibility for the consequences of their actions. The political subject of sabotage simply takes this attribute of the system as given and turns it against the system itself, thereby becoming a constituent (not the author) of an exacerbated relation. This is what it means to act today.

Why linger over a discredited category such as sabotage? It is not to suggest that sabotage is what people ought to do, or that they ought to use that word to name what they are already doing. The conventional association of sabotage with destructiveness and criminality runs too deep to recommend such provocations, and, in any case, the account presented above would imply there is no need to promote sabotage because sabotage is already under way. This is precisely the value of highlighting it as a category. The veritable evacuation of liberal democratic institutions and procedures as venues for effective political action, the hyperobjective qualities of climate change, and the apparently intractable character of the petrocultures that subtend both have generated a social condition that is experienced by many as postpolitical, in which the meaning of action is difficult to discern. However, politics is a property of the world that cannot be eradicated so easily. The task of a critical theory of petroculture is to identity the forms that politics does and might take under conditions that would otherwise seem to erase it, forms that are consistent with the material realities of that culture. The intuition sketched briefly here is that sabotage might be one of those forms, particularly in light of the qualities attributed to it in the latter parts of this chapter—its *internality* to systems, its *mediational* and *postsubjective* character—which exceed the potential attributed to it in Mitchell's account of carbon democracy. It bears remembering that Mitchell's saboteurs did not disrupt the movement of coal because they wanted to end the extraction and consumption of coal. They did it to accomplish a redistribution of the value generated by the carbon economy, a demand whose satisfaction was predicated on the continuity of that economy. The same goes for the sabotage carried out by striking refinery workers in France today.[39] There is nothing wrong with this sort of action, and it can result in many good things, but it is not a type of sabotage oriented to enacting a future social order "after oil."[40] Who and where, we might ask, are the actors internal to petrocapitalism who will pull the "fine thread of deviation" and unleash a saboteurial unraveling for which there is no effective countermeasure? Which confederacies will be established and mobilized between human and inhuman agencies, and what will be the intimate knowledges by which their mediation is accomplished? And who, or what, will be the saboteurial political actor that, in its constant activity, does not appear and never makes a speech or a demand, and whose action cannot be traced to its source? What and where is the invisible organization, and what forms of mediation will it materialize? These are questions that might orient us to the enigma of what it means to act into the apparent impasse of today's energy culture.

Notes

1. Paolo Virno, "Virtuosity and Revolution: The Political Theory of Exodus," in *Radical Thought in Italy: A Potential Politics*, edited by Paolo Virno and Michael Hardt (Minneapolis: University of Minnesota Press, 1996), 189.
2. Timothy Morton, *Hyperobjects: Philosophy and Ecology after the End of the World* (Minneapolis: University of Minnesota Press, 2013), 1.
3. Virno, "Virtuosity and Revolution," 189.
4. Timothy Mitchell, *Carbon Democracy: Political Power in the Age of Oil* (New York: Verso, 2011), 8.
5. Mitchell, *Carbon Democracy*, 19.
6. Mitchell, *Carbon Democracy*, 27.
7. Mitchell, *Carbon Democracy*, 252.
8. See Pierre Dubois, *Sabotage in Industry*, translated by Rosemary Sheed (New York: Penguin, 1979).
9. Arturo Giovannitti, "Introduction to Pouget's Sabotage," in Émile Pouget, *Sabotage*, translated by Arturo M. Giovannitti (Chicago: Charles H. Kerr & Co., 1913 [1912]).
10. Elizabeth Gurley Flynn, "Sabotage: The Conscious Withdrawal of the Workers' Industrial Efficiency," in *Direct Action and Sabotage: Three Classic Texts from the 1910s*, edited by Salvatore Salerno (Oakland, CA: PM Press, 2013), 91.
11. See US Supreme Court, *Burns v. United States*, 274 U.S. 328 (1927), argued 24 November 1926, decided 16 May 1927, https://supreme.justia.com/cases/federal/us/274/328/case.html.
12. Flynn, "Sabotage," 98. As Salerno recounts, Boyd would later renounce sabotage in exchange for a pardon, leading Flynn to append a note to her pamphlet decrying his "cowardice" and to speculate that he might have been a provocateur. In 1917, Flynn herself would renounce sabotage in an effort to secure Woodrow Wilson's intervention in charges brought against her under the Espionage Act. See Salvatore Salerno, "Introduction," in *Direct Action and Sabotage: Three Classic Texts from the 1910s*, 17–18.
13. William E. Trautmann, "Direct Action and Sabotage," in *Direct Action and Sabotage: Three Classic Texts from the 1910s*, 42. See also Pouget, *Sabotage*, 55.
14. Thorstein Veblen, *The Engineers and the Price System* (Kitchener, ON: Batoche Books, 2001), 7.
15. On planned obsolescence, see Giles Slade, *Made to Break: Technology and Obsolescence in America* (Cambridge, MA: Harvard University Press, 2007). In 2015, the US Environmental Protection Agency sanctioned Volkswagen for installing "defeat devices" in its diesel vehicles in order to circumvent emissions standards, a practice later revealed to be relatively common in the industry. See Megan Geuss, "Volkswagen's Emissions Cheating Scandal Had a Long, Complicated History," *arsTECHNICA*, 24 September 2017, www.arstechnica.com.
16. On publicity, see Darin Barney, "Publics without Politics: Surplus Publicity as Depoliticization," in *Publicity and the Canadian State: Critical Communications Approaches*, edited by Kirsten Kozolanka (Toronto: University of Toronto Press, 2014), 72–88.

17. See Darin Barney, "We Shall Not Be Moved: On the Politics of Immobility," in *Theories of the Mobile Internet: Materialities and Imaginaries*, edited by Jan Hadlaw, Thom Swiss, and Andrew Herman (New York: Routledge, 2014), 15–24.
18. Evan Calder Williams, "Manual Override," *New Inquiry* 50 (21 March 2016), https://thenewinquiry.com/manual-override/.
19. Laura Kane, "Indigenous Protesters Build Homes in Trans Mountain Pipeline's Path," *Globe and Mail*, 7 September 2017, https://www.theglobeandmail.com/news/british-columbia/indigenous-protesters-build-homes-in-trans-mountain-pipelines-path/article36207510/.
20. On logistics and politics, see Deborah Cowen, *The Deadly Life of Logistics: Mapping Violence in Global Trade* (Minneapolis: University of Minnesota Press, 2014); and Ned Rossiter, *Software, Infrastructure, Labor: A Media Theory of Logistical Nightmares* (New York: Routledge, 2016).
21. See John Durham Peters, *The Marvelous Clouds: Towards a Philosophy of Elemental Media* (Chicago: University of Chicago Press, 2015).
22. On the role of labor in the transition from hydraulics to coal power, see Andreas Malm, *Fossil Capital: The Rise of Steam Power and the Roots of Global Warming* (London: Verso, 2016).
23. Williams, "Manual Override."
24. Williams, "Manual Override."
25. Williams, "Manual Override."
26. Williams, "Manual Override, Lecture 2: The Sabotage of Time," lecture, Center for Transformative Media, Parsons The New School for Design, New York, 10 March 2014.
27. On the multiple material agencies at work in pipeline developments, see Andrew Barry, *Material Politics: Disputes along the Pipeline* (London: Wiley-Blackwell, 2013).
28. Williams, *Manual Override: The Sabotage of Capital*. 2013-14 Fellows Lectures. Centre for Transformative Media, Parsons: The New School for Social Design. This quotation is taken from the abstract posted at http://ctm.parsons.edu/events/lecture-series/2013–14-fellows-lectures/.
29. Richard Grusin, "Radical Mediation," *Critical Inquiry* 42 (Autumn 2015): 130.
30. Grusin, "Radical Mediation," 129.
31. Williams, "Manual Override, Lecture 1: The Sabotage of Production," lecture, Center for Transformative Media, Parsons The New School for Design, New York, 4 December 2013.
32. On strikes and sabotage, Dubois observes: "A strike only constitutes sabotage in the firm where it takes place . . . as a general phenomenon, the strike is only a temporary form of sabotage." Dubois, *Sabotage in Industry*, 37.
33. Stefano Harney and Fred Moten, *The Undercommons: Fugitive Planning and Black Study* (Brooklyn, NY: Minor Compositions/Autonomedia, 2013), 74–75.
34. Hannah Arendt, *The Human Condition* (Chicago: University of Chicago Press, 1958), 180–81.
35. Williams, "Manual Override, Lecture 1."
36. Williams, "Manual Override, Lecture 1."
37. Williams, "Manual Override, Lecture 2."
38. CBC News, "Heinz Closes Leamington Plant, 740 People Out of Work," *CBC News*,

14 November 2013, http://www.cbc.ca/news/canada/windsor/heinz-closes-leamington-plant-740-people-out-of-work-1.2426608.

39. Emmanuel Jarry and Ingrid Melander, “Pilots, Oil Workers Strike as France Seeks Way Out of Crisis,” *Reuters*, 30 May 2016, https://www.reuters.com/article/us-france-politics-protests-idUSKCN0YL1PM.
40. Petrocultures Research Group, *After Oil* (Morgantown: West Virginia University Press, 2016).

CHAPTER 17

Strike

Antonio Negri

Translated by Evan Calder Williams

In the beginning, there was sabotage: throwing the *sabot*, the wooden clog, into the gears of the machine—to pay back the desperation of exploitation, fatigue, and pain. Beyond expressing desperation, however, sabotage expressed intelligence: it revealed a primordial fact, that without "living labor," there would be no production, that workers' refusal stopped production, negated it, that capitalist production was based not just on machinery but also on obedience—that, therefore, the relation with the boss happened through his (the boss's) mechanical prostheses. This had not happened with agrarian labor or the regime of feudal property: the exploitation of the agrarian worker, the peasant, was external to mechanical relations, and the standards governing the distribution of ground rent were constructed in a manner external to any mechanical measure. With the birth of manufacturing and the development of large-scale industry, human labor was instead dragged inside the machine. The standards of exploitation and of distribution were administered *inside* the system of machines. Protest, resistance, and refusal of exploitation therefore could only be given within and *against* the system of machines. Sabotage is the spark of this protest. Often individual. But sabotage can also be collective (perhaps it's always collective), because it's hard to imagine a spark without a fire that burns. In fact, the strike can arise, as a collective behavior, when the spark, through organization and the awareness of the organization of labor, succeeds in setting fire to the prairie. The strike therefore opens a new epoch in the history of the relation between classes.[1] A conflictual relation like this, on a terrain marked by manifest and continual action between two classes in struggle, had never been seen: not between slave masters and enslaved populations, not between nobility and bourgeoisie . . . the other examples are lacking. The strike, therefore, appeared for the first time in history as a rupture internal to the relation of class subjects in struggle—who cannot resign themselves, one giving in to the other. *Aventino* is impossible.[2] That "contractual" relation, which in capitalism is established between the boss and the worker, represents the recognition of the worker as inside the productive system (as part of capital, as "variable capital," living labor that gives value to the commodities produced). This recognition is necessary for the capitalist. Never had labor been so internal to its command, and never had it been so integrated into profit. But this contractual relation is falsified by the fact that the boss doesn't bargain with the worker (with the mass of workers) in good faith, evenly dividing up the profits—but does so in an unequal manner, determining the price of labor-power (and its reproduction) in an underhanded and asymmetrical way. With the strike—that is, negating itself as labor-power—the worker class disavows the validity of the contract, bringing to light the asymmetry and injustice. In this way, the strike thereby presents an extraordinary ambivalence: it can be a form in which the labor contract is organized; it can, however, also be an expression of worker power, of worker refusal to be enclosed in the contractual cage. Normally, it is somewhere between these things. As Rosa Luxemburg said, "the struggle against the reduction of money-wages also signifies struggle against the market character of labor-power, that is, against capitalist production taken in full. The struggle against the drop of money-wages is no longer a battle on the terrain of the market economy but a revolutionary attack against the foundations of this economy; it is the socialist movement of the proletariat."[3] A struggle over wages, over the distribution of profits gained through production, but also a struggle beyond distribution, a revolutionary struggle to seize power over the productive machinery: it's in this way that we should read Rosa's definition of the strike.[4] Technically, the strike in large-scale industry is essentially defined as a rupture in, and recomposition of, the labor relation: but it is something else entirely, perhaps even the opposite. It is clear to see, then, why bourgeois law and its civil codes worked so hard not only to recognize the strike but also to accept its use. Previously, the French Revolution had not wanted to include the strike among the rights of the working man—indeed, it banned it. It would take years and years of struggle for the strike to be admitted into constitutional rights. And it would take plenty of the dead—killed by the state's police or the bosses' Pinkerton men, by layoffs and the subsequent misery of workers and their families, years of jail and Mafia intimidation, free-market emigration and fascist massacres. So the strike, through these events, became a right. There isn't constitutional democracy without the right to strike. *Gloria*, therefore, *in excelsis coelis* for the right to strike? Yes and no. In fact, when the strike became a right of massified labor-power in the democracies of the *souche* West, the control of labor-power then passed through other instruments. That of the union will be fundamental. The union is born as an association for the conquest, definition, and development of the rights of workers. It comes to be an essential instrument for the emancipation of labor and the very basis for the construction of parties that fight for workers and the oppressed. However, the union is marked by the same ambiguity as the strike: it can be both an instrument for bargaining and an instrument for breaking the capitalist relation. In the history of industrial relations, this duplicity was often seen and expressed, and it generated major contradictions and battles. For long periods, anarcho-syndicalism and/or revolutionary syndicalism developed wage demands through power struggles. In the twentieth century, faced with new and increasingly strong crises of capitalist development (and the crises of the 1920s, in particular), several political forces of more advanced capitalism had thus thought of and operated toward a participatory integration of the union into the direction of development. Keynesianism as economic theory and the Rooseveltian New Deal as capitalist *governance* achieved this reformist formulation with real success. In this situation, the strike was—so to speak—bureaucratized, and in this condition, many protests and radical ruptures between the base and top of the unions emerged. However, this situation was to last only long enough to permit the bosses' class a total reform of industrial relations. In 1973, the authors of the *Tricontinental* report denounced the fact that, since '68, in factories and society alike capitalists had to submit to the diktat "of workers, students, women, and blacks": here, other instruments had to function when the exercise of the strike had taken to pieces the assizes of production and when repression of the strike no longer had effect—new instruments, therefore, had to be invented. The working class is then expelled one by one from the factories through the automation of the plants; the informatization of society permits a larger spatial mobility and temporal flexibility of production and, in consequence, the precarization of large shares of labor-power; the feminization of labor enormously enlarges the labor market; and finally, information technology will permit the expansion and diffuse control of the socialization of labor. What form of strike for this new class of workers, for this multitude of citizens submitted to capitalist domination? If the domination of capital extends itself over society, it's in society that the organization of the strike, of the rupture in the relation of domination, must today be given and defined. And if capitalist command has now become a biopower, it is in the world of life, in its entire extension, that the rupture of the relation of exploitation will be redefined and eventually organized. Today, the strike recognizes itself as essentially a form of the refusal of labor, of a blockage in the social machine of production, and of confrontation with the capitalist state. We are midway along the path.[5]

Notes

1. [The wording is of import here: Negri does not write "the history of class relations" (or "the history of the class relation") but "the history of the relation between classes." This indicates a continued emphasis on the idea developed in the lectures on the *Grundrisse*, for instance, that there are two subjects of capital.—Trans.]
2. [His reference is to the Aventine Secession, a reference that is itself double. The more immediate echo is the boycott by and withdrawal of the Italian Socialist Party from the Chamber of Deputies, following the murder of Socialist politician Giacomo Matteotti by fascists in 1924. The boycott was intended to spur King Victor Emmanuel III into dismissing Mussolini. The tactic failed: by 1925, Mussolini banned the Socialist party from returning and declared a dictatorship. This sequence of events was named after the Aventine Secession of 494 BCE, when the plebeians seceded from Rome (and camped on the Mons Sacer) to force a negotiation with the patricians. The end result was the emergence of the tribunes, plebeian magistrates who would hypothetically represent pleb interests—and, in terms of this essay, a "giving in" of one side to the other in the form of an inclusion into a putatively more egalitarian contract.—Trans.]
3. [Rosa Luxemburg, *Einführung in die Nationalökonomie*, in *Gesammelte Werke*, vol. 5 (Berlin: Dietz Verlag, 1990), 524–778.—Trans.]
4. [Luxemburg is specifying the money wage (*salario relativo*) in distinction to real wages.—Trans.]
5. [Negri's apparent reference here is to the first lines of Dante's *Divine Comedy*: "Nel mezzo del cammin di nostra vita" ("Midway on our life's journey" in the Pinsky translation, or "Midway on the path of our life").—Trans.]

CHAPTER 18

Energized Antagonisms: Thinking beyond "Energy Culture"

Matthew Huber

In 1943, anthropologist Leslie White published the article, "Energy and the Evolution of Culture." He began with the astounding pronouncement that "everything in the universe may be described in terms of energy."[1] At a time when social scientists were keen to demonstrate their scientific credentials, he postulated a series of laws of energy and culture. For instance: "We have . . . *the* law of cultural evolution: culture develops when the amount of energy harnessed by man per capita per year is increased."[2] These laws emerged from a staunchly materialist analysis in which the essential basis was the fact that humankind must produce the material conditions of its existence—food, shelter, and other necessities. The less energy devoted to this work of material existence, the more time and space can be devoted to what White calls "culture building."[3] Culture building generally means cultural-symbolic systems—writing, religion, the arts, laws, and rituals. But this raises the question: what is culture? Is culture a word for differentiated communities across the planet? Is it a name for art, music, and aesthetic production? Is it a "level" of social life—a level of ideas or beliefs? Is it life itself material practice?

White was dealing with a rather narrow definition of culture. For White, it was something akin to what Eurocentric colonialists call "civilization"—a modernist marker for advancement and progress. Culture is a confusing and troubled concept. Raymond Williams famously said it is "one of the two or three most complicated words in the English language."[4] I'm a geographer, and geographers have thought a lot about culture and have long criticized the concept. In 1980, James Duncan accused cultural geographers of using "culture" as an explanatory concept with "ontological status and causative power."[5] For many geographers, culture was reified as a holistic force that caused individuals to behave in certain ways. This leads to what we might call "culturalism" as a form of cultural determinism where inequality is explained as an inherent and unchangeable aspect of someone's "culture." Moreover, Duncan critiques the ways in which geographers assume "internal homogeneity" within a specific concept of a culture—unified by a common set of practices, ways of living, belief systems, and so on.[6] I think that our notions of "energy culture" are plagued with this vision of "internal homogeneity," a point I will make clear over the course of this chapter.

Don Mitchell argues more provocatively that "there is no such thing as culture."[7] He means that literally—there is no "thing" we can hold and grasp that we can name as culture. We too often reify culture as a thing that has its own powers. Mitchell suggests instead that culture is a powerful idea that "has developed under specific historical conditions and was later broadened as a means of explaining material differences, social order and relations of power."[8] He suggests we should dispense with the study of a thing called culture and instead seek to understand "how the idea of culture (rather than culture 'itself') has been deployed by powerful social actors."[9]

Despite this conceptual confusion and political problematics, culture has become one of the primary concepts through which we seek to understand energy at present. For this emerging body of work,[10] culture seems to invoke what Raymond Williams defined as "as a whole way of life."[11] This notion of culture as a way of life and its imbrication with energy leads us to consider the energy-intensive forms of life predominant in the industrialized North. In the introduction to *Oil Culture*, Ross Barrett and Daniel Worden argue that "oil culture . . . has helped to establish oil as a deeply entrenched way of life in North America and Europe by tying petroleum use to fundamental sociopolitical assumptions and aspirations, inventing and promoting new forms of social practice premised on cheap energy."[12] In this view, energy becomes a *unifying force*, a kind of material glue that ties together and becomes the condition of possibility for a whole way of life or culture. Without energy, this culture could not be possible. Energy makes possible a culture of family, freedom, and the American way.[13]

In this chapter, however, I propose a different view of energy, one that sees it not as a unifying force that creates a whole culture, but rather as a material force of division and antagonism: energy as the basis for social struggle between groups and classes within society, and as a material basis for the accumulation of wealth and power by one specific group at the expense of another. This is another way of saying we need to see energy not as fueling some unified idea of a "culture," but as fueling class struggles and inequality.

In one sense, I'm following Dominic Boyer's call for us to look at "energopolitics" or "energopower"—a conception of modern social power's constitutive relation with energy—though I do not figure power in such a dispersed and Foucauldian register.[14] I'm advocating an energy-power analytic that sees conflict and struggle between social groups, fractions of capital, the state, and other institutions at the heart of the analysis (in other words, a more Gramscian energopower analysis). Common ways of thinking about energy in this way—focused on material division and antagonism—already exist. Oil is seen as the basis for state power,[15] "Big Oil," and corporate power,[16] and as the basis of US and other forms of imperialism.[17] These energy-power analytics focus on access and control over the *production* of energy resources. For some reason, we lose this sense of antagonism when we consider the landscapes of energy consumption. Energy consumption is too often conceptualized as creating a singular, monolithic, and homogeneous "culture." This is the problem I confront in this chapter: how can we think of energy consumption in terms of class divisions, inequality, and antagonisms?

This critique of energy culture is broken up into two specific ideas of culture. The first concerns the culture of suburbia. I aim to retell the emergence of postwar suburbanization as an oil- and electricity-powered class formation based on privileges rooted in race, gender, class, and property. I will illustrate this through the case of the second Levittown development, outside Philadelphia. The second concerns a culture of excess. I will examine some discourse surrounding "peak oil" and its conception of an *excessive* energy culture. I argue that this vision of an excessive energy culture comes from a limited, middle- and upper-class perspective and fails to take into account larger realities of economic insecurity and material deprivation. While a politics of excess implies a politics of limits and "less," in this age of debt, wage stagnation, increasing poverty, and inequality, we need an energy politics focused on abundance and "more." These two visions of culture are obviously intertwined, but I think a focus on one or the other leads to very different ideas of political struggle. Moreover, this chapter focuses on a very narrow version of "energy culture"—that of high-energy suburban consumers. Obviously, a more expansive critique of energy culture would need to also examine non-Western forms of energy-society relations.

Energy and Suburban Culture

Insofar as oil powers private transportation (automobility), it also powers the dispersed privatized geographies of single-family suburban home ownership.

The idea of a singular and homogeneous suburban culture powered by oil was well understood by the petroleum industry in the postwar era. Oil advertisements during the 1950s consistently depicted a homogeneous vision of white, heteronormative, middle- and upper-class suburban consumers of petroleum products.[18] For example, Shell issued an ad campaign called "From A to Z—An Alphabet of Good Things about Petroleum." One ad claimed "M is for Mother," depicting a succession of practices, each one dependent upon oil in some way. "Things like wash and wear fabrics and stain resistant paints are making home routine jobs easier all the time. Who's responsible for so many of those time saving products? Mother's little helper—oil!" This made Mother's life appear much easier because of oil, obscuring the highly gendered dynamics of postwar suburbia. Privatizing social reproduction into the hands of a single housewife actually created, in Ruth Cowan's phrase, "more work for Mother."[19] Overall, the oil industry projected the idea featured in an Esso ad from 1953. The ad features the typical white, middle-class family in a car and proclaims: "Petroleum helps to build a better life." In this case, "petroleum" means not only the gasoline powering the car, but also the oil-based lubricants, synthetic tires, and, indeed, the road itself that is built from petroleum-based bitumen or asphalt. This creates a narrative I have called "the unavoidability of oil."[20] Life—and culture—becomes unimaginable without it.

But what really built this culture—this culture of higher wages, of home ownership in the suburbs? Was it oil? To understand that, you need to look back to the 1930s and understand how this oil culture was created out of class struggle. A labor historian described the conditions in the era: "In 1934 labor erupted. There were 1856 work stoppages involving 1,470,000 workers, by far the highest count . . . in many years."[21] Political economist Robert Brenner explains how this eruption caused the very significant labor victories of the New Deal: "It was this explosion of mass direct action outside the electoral-legislative arena that constituted the indispensable precondition for the popular gains of the New Deal."[22]

The National Labor Relations Act of 1935; the creation of the Federal Housing Administration, making home ownership more affordable; the creation of the welfare state, social security, and unemployment insurance: all this allowed for the explosion of postwar suburbanization. But these were not the benevolent good deeds of an enlightened President Roosevelt. These policies were the result of mass popular protest and pressure. They did create a landscape powered by cheap oil, but it was a landscape forged through social struggle. Moreover, these victories were very partial and based on exclusion.

These popular gains were based on entrenched sets of privilege involving race, gender, class, and, perhaps above all, property ownership. From 1934 to 1962, the federal government underwrote $120 billion in new mortgages, and an astonishing 98 percent of this went to white families.[23] These real estate policies were based on racialized redlining practices that made entire inner-city neighborhoods off limits for mortgage lenders. A geography of dispersed suburban private home ownership was made possible by oil, but that culture should not stand in for "American" culture. This was an exclusionary geography of privatism that also wrapped up property values with race. As Thomas Sugrue puts it, "in postwar America, where you lived shaped your educational options, your access to jobs, and your quality of life. The purchase of a home also provided most Americans with their only substantial financial asset. Real estate was the most important vehicle for the accumulation of wealth."[24]

I will now examine the specific case of Levittown, Pennsylvania—the second Levittown after the first on Long Island. William Levitt was quoted on several occasions as wanting to create all-white communities, and he expressly argued that opening up the communities to African Americans would lead to an inability to sell the homes or would depress property values.[25] The problem was that this was illegal because of a Supreme Court ruling that banned racialized housing covenants, which were quite popular from the 1920s to the 1940s. This ruling came at the embarrassingly late date of 1948.[26] The Levitts were the object of campaigns by the National Association for the Advancement of Colored People (NAACP), which pressured the governor of Pennsylvania to prosecute the Levitts for housing discrimination.[27] The governor admitted he was against this kind of blatant housing discrimination, but said nevertheless he did not want "to impose conditions of occupancy on a private builder."[28]

Meanwhile, a nearby Quaker community, whose members were committed peace and civil rights activists, began a many-year campaign to recruit African American homebuyers to integrate Levittown.[29] A class bias led such efforts to recruit a so-called "respectable" middle-class African American family to move to Levittown and integrate it. They found such a family in William and Daisey Myers. Sugrue describes them:

> Both in their mid-thirties, William was a World War II veteran and a graduate of the Hampton Institute who worked as a refrigeration technician in nearby Trenton, New Jersey, and Daisy was a schoolteacher and a member of the Bristol Township recreation board. Like most Levittown families, they were baby-boom parents, with a newborn, a three-year-old,

> and a five-year-old. But however much they resembled their white neighbors, when word got out that they had purchased the Levittown house, a grassroots mobilization against them began.[30]

Sugrue describes several forms of resistance against the Myerses: a petition against them signed by thousands of Levittowners; protests by hundreds of people who came out on the streets to challenge the home purchase; and attacks on their house, which included the breaking of windows.[31] A cross was burned, and "KKK" was written on the property of a neighbor who supported their move into the neighborhood. Local Republican leader James H. Paul blamed the Myers family for the violence and riots, and told them to "go back where you came from."[32] This outrage and privilege were largely expressed as racism, but also as a self-interested and rational consumer defense of property values. A white Levittowner named J. P. Walsh framed it this way: "We feel we have the moral right to protect our property values and rear our children in the best possible surroundings."[33] One neighbor candidly told a reporter that Bill Myers "is probably a nice guy, but every time I look at him I see $2000 drop off the value of my house."[34] It should be noted that hundreds of left-leaning, civil rights–supporting white activists came out for the Myerses,[35] providing them with moral support and financial assistance. Despite this aid, the Myers family moved out of Levittown two years later.[36]

This geography of privatism and privilege became sedimented into the landscape, creating separate and unequal geographies of affluence and poverty, opportunity and misery. Oil did not power an entire culture, but it became the material basis for an exclusionary geography of privilege and privatism. White, middle-class Americans were consuming massive amounts of energy and living prosperous lives, while the inner cities of America crumbled. This brings to mind a 1970 poem by poet and musician Gil Scott-Heron, which distilled this inequality in one single absurdity: "A rat done bit my sister Nell (with Whitey on the moon)."[37] Although Heron directed his disdain toward perhaps the most ludicrous use of societal resources, to explore space, he might as well have also focused on "Whitey" in the suburbs. Both on the moon and in the suburbs, white people were using tremendous amounts of energy—the energy of rocketships, the energy of automobiles—to take themselves far away and seclude themselves from earthly difference. Ultimately, Heron's critique is of a society that uses and wastes so much energy on behalf of the privileged and white, while black America suffers with a despicable lack of energy. "No hot water, no toilets, no lights, / But Whitey's on the Moon."[38]

The inherent inequalities of this new form of housing segregation exploded in the 1960s with widespread revolts throughout the inner cities of the United States: the Watts area of Los Angeles, Baltimore, Detroit, and so on. The revolts of urban African Americans in the 1960s were fundamentally about these stark inequalities—enormous material prosperity for white suburban America, while the cities and public housing projects burned. These were revolts against the postwar accord between the state, labor, and upper-middle-class white America for constructing an exclusive geography of energized wealth and privilege. These were revolts among those still forced to rely on decrepit public transit systems to get to any employment. These were revolts against the hollowing out of the tax base and basic social services in urban communities across the country.

Thus, oil did power a culture. But we often generalize this white, middle-class "culture" as if it stands in for the entire society. What oil provided was the energy and the power to sever oneself from society. It energized an atomization of social life. Suddenly, with the power to travel in a private car to one's private home filled with electrified diversions, one had no need to engage with the idea of "the public." Energy powered a geography of profound isolation. As Guy Debord put it, "the reigning economic system is a *vicious cycle of isolation*. Its technologies are based on isolation, and they contribute to that same isolation. From automobiles to televisions, the goods that the spectacular system *chooses to produce* also serve it as weapons for constantly reinforcing the conditions that engender 'lonely crowds.'"[39] This geography of privatism, powered by oil, grew to become a political force shaping our politics. The basis of this politics is a disdain for even the notion of society. As one letter to President Richard Nixon from a white suburban father in Charlotte, North Carolina, put it succinctly: "I have never asked what anyone in government or this country could do for me; but rather kept my mouth shut, paid my taxes and basically asked to be left alone."[40] As US politics shifted rightward, and as the disdain for taxes, wealth redistribution, and government programs intensified, these white homeowners forgot that suburban life is itself a product of public investments—public mortgage programs, public investment in highways, and public education like the GI Bill. But as disdain for the idea of society and the public grew, it not only deepened the divisions and poverty of those left out of this New Deal project—African Americans, migrant workers, and so on—but also served to erode the incomes and lives of those middle-class suburban homeowners themselves. Unions were busted. Wages were cut. Government support was slashed.

So oil did not power a unified and monolithic culture. Rather, it powered a very partial and narrow political project based on what Evan MacKenzie calls an "ideology of hostile privatism."[41] This ideology of privatism has created vast inequality and material deprivation; yet many analysts of energy culture view American culture as one of "excess."

Energy and a Culture of Excess

I now turn to the ways in which peak oil discourse constructs "energy culture." In particular, I highlight their focus on a culture of *excess*. This construction of an *excessive* energy culture is rooted in the idea that fossil fuels themselves are a onetime "excess" of millions of years of sunlight laid down by nature. This creates the conditions for what Richard Heinberg calls a "party." Heinberg writes, "Through the one-time-only process of extracting and burning hundreds of millions of years' worth of chemically stored sunlight, we built what appeared (for a brief, shining moment) to be a perpetual-growth machine. We learned to take what was in fact an extraordinary situation for granted. It became normal."[42]

With this bounty it is assumed that American energy culture is one of excess, waste, and overconsumption. Another famous peak oil icon, James Howard Kunstler, describes American culture as such: "We've become a nation of overfed clowns and crybabies, afraid of the truth, indifferent to the common good, hardly even a common culture, selfish, belligerent, narcissistic whiners seeking every means possible to live outside a reality-based community."[43] To be sure, there is a small slice of our energy culture that does consume too much: the top 10 percent of society and the superrich. However, overall, does the American economy look like one based on excess and overconsumption? A 2016 survey found that 63 percent of Americans could not afford a $500 emergency expenditure for a health care issue or broken-down car.[44] One-quarter of Americans working forty hours a week or more are living below the poverty line.[45] This is not an economy of excess, even if it takes significant energy for these workers to reproduce their everyday lives.[46] Just because they consume a lot of energy does not mean they feel rich or like "overfed crybabies." The surface of energy profligacy obscures an enormous amount of economic insecurity and economic deprivation. In fact, recent studies have revealed that the new centers of poverty in the United States are in the suburbs where one needs to consume a lot of energy just to live.[47] Consuming lots of energy to reproduce an impoverished life is the new normal in the suburbs.

For most peak oil discourse, the solution involves what I'm calling a "politics of less." Richard Heinberg, in his book whose very title, *Powerdown*, implies a politics of less, wrote a whole chapter calling for "self-limitation." He ends it with these words:

> The way out is to restrict per-capita resource usage and to reduce the human population. If we take the Powerdown path, then alternative energy sources could help. If we refuse to power down, then nothing will help. In the end, self-limitation is the only answer that counts, but that is the answer that no one wants to hear. So we sit, and wait, and assume, and deny. And as we wait, the signs of depletion worsen and global resource wars loom. If we refuse to take the hard Powerdown path, after a while we will simply have no choice: we will compete for what is left . . . or we will die.[48]

This is a bleak vision, indeed, and it's hard to understand how it might appeal to a broad popular base. Of course, the peak oil narrative itself mainly appeals to a relatively affluent, middle- or upper-class consumer who is relatively well off and *feels* excessive.[49] Peak oil advocates generalize their own energy-intensive lifestyle with an entire culture of ordinary "middle class" Americans. But most Americans are not feeling particularly excessive right now. And a political vision of "limits," "powerdown," and "less" is not likely to appeal to a struggling working class drowning in debt.

A peak oil advocate might reply that popular appeal is irrelevant when impending scarcity is reality. Nevertheless, it is worth pointing out that the entire cottage industry of peak oil literature has been silenced by a torrent of cheap, fracked tight oil in North Dakota, Texas, and elsewhere. Prices have collapsed amid the glut. There is simply too much oil right now.[50] From the perspective of climate change, there is too much oil left to burn. Regardless of where you stand on peak oil—it will peak someday—we need a different language for energy politics, one that emphasizes a politics not of scarcity but of abundance. Economic geographer Erica Shoenberger outlines how our environmental crisis needs to create a politics of more. She says we are being "self-involved" when we only focus on the wasteful consumption patterns of a minority of profligate Americans.[51] The majority of Americans and, of course, the majority of the world do not live excessive, wasteful lives. Schoenberger says, "Much of the rest of the world's population needs to consume more. They need more food, more clean water, more sanitation, more electricity, more industry."[52]

A politics of more simply means adjusting to new energy possibilities. As Andreas Malm's work makes clear, we are in the midst of a transition: from a "stock" energy resource of fossil fuels that are finite and when combusted are destroyed forever, to the "flow" energy resources of sun, wind, and tides that are inexhaustible and abundant. The material basis of renewables provides an escape from the politics of less. As Malm points out, the real challenge for capitalism is finding ways to profit and control wealth accumulation from such flow resources.[53] If our renewable energy transition were managed by the public sector—much like the 1930s Tennessee Valley Authority or the Rural Electrification Administration—we could build a public infrastructure for cheap, abundant energy for all. The public crisis of climate change demands such a public-sector approach. Such a program cannot be won by focusing on scarcity and "powering down" a culture of excess. It must offer more and cheaper energy to precarious populations.

Conclusion

We need to think beyond "energy culture." In reality, energy is not something that powers a singular, homogeneous culture. It powers social power.[54] It's the material basis for some parts of society to accumulate wealth and power at the expense of others. We see how suburban energy culture was about powering a certain kind of propertied privilege based on race, class, and gender. We see that the narrative of a culture of "excessive" energy consumption also obscures class divisions and economic insecurity (that our energy politics must answer to). Our energy systems could be seen as the material basis for abundance and security for all. This is actually what Marx believed when he said:

> [The capitalist] is fanatically intent on the valorization of value; consequently, he ruthlessly forces the human race to produce for production's sake. In this way, he spurs on the development of society's productive forces and the creation of those material conditions of production which alone can form the real basis of a higher form of society, a society in which the full and free development of every individual forms the ruling principle.[55]

Fossil fuels could be seen as the material basis for an inexhaustible and sustainable solar energy society. Fossil fuels need not only be seen as the original sin of industrialism that we must reject and power down from, but as a dirty springboard to an abundant and clean energy future. But that would take a massive political struggle over who controls our energy systems. It would be about deploying energy not for profit, but for the human and ecological needs of society and future generations.

Notes

1. Leslie White, "Energy and the Evolution of Culture," *American Anthropologist* 45, no. 3 (1943): 335.
2. White, "Energy and the Evolution of Culture," 338.
3. White, "Energy and the Evolution of Culture," 339.
4. Raymond Williams, *Keywords: A Vocabulary of Culture and Society* (New York: Oxford University Press, 1976), 83.
5. James S. Duncan, "The Superorganic in American Cultural Geography," *Annals of the Association of American Geographers* 70, no. 2 (1981): 181.
6. Duncan, "The Superorganic in American Cultural Geography," 181.
7. Don Mitchell, "There's No Such Thing as Culture: Towards a Reconceptualization of the Idea of Culture in Geography," *Transactions of the Institute of British Geographers* 20, no. 1 (1995): 102–16.
8. Mitchell, "There's No Such Thing as Culture," 103.
9. Mitchell, "There's No Such Thing as Culture," 104.
10. Janet Stephenson, Barry Barton, Gerry Carrington, Daniel Gnoth, Rob Lawson, and Paul Thorsnes, "Energy Cultures: A Framework for Understanding Energy Behaviours," *Energy Policy* 38, no. 10 (2010): 6120–29; Sarah Strauss, Stephanie Rupp, and Thomas Love, eds., *Cultures of Energy: Power, Practices, Technologies* (London: Routledge, 2013); Stephanie LeMenager, *Living Oil: Petroleum Culture in the American Century* (New York: Oxford University Press, 2013); Ross Barrett and Daniel Worden, eds., *Oil Culture* (Minneapolis: University of Minnesota Press, 2014).
11. Williams, *Keywords*, 13.
12. Barrett and Worden, *Oil Culture*, xxv.
13. Matthew T. Huber, *Lifeblood: Oil, Freedom, and the Forces of Capital* (Minneapolis: University of Minnesota Press, 2013).
14. Dominic Boyer, "Energopolitics and the Anthropology of Energy," *Anthropology News* 52, no. 5 (2011): 5–7.
15. See, for example, Fernando Coronil, *The Magical State: Nature, Money, and Modernity in Venezuela* (Chicago: University of Chicago Press, 1997); and Robert Vitalis, *America's Kingdom: Mythmaking on the Saudi Oil Frontier* (Palo Alto, CA: Stanford University Press, 2007).
16. See, for example, Steve Coll, *Private Empire: Exxon-Mobil and American Power* (London: Penguin, 2012).
17. See, for example, Greg Muttitt, *Fuel on the Fire: Oil and Politics in Occupied Iraq* (New York: New Press, 2012).
18. Huber, *Lifeblood*, 73–95.
19. Ruth Cowan, *More Work for Mother: The Ironies of Household Technology from the Open Hearth to the Microwave* (New York: Basic Books, 1983).

20. Huber, *Lifeblood*, x.
21. Irving Bernstein, *Turbulent Years: A History of the American Worker, 1933–1941* (Boston: Houghton Mifflin, 1971), 217.
22. Robert Brenner, "Structure versus Conjuncture: The 2006 Elections and the Rightward Shift," *New Left Review* 43 (2007): 38.
23. George Lipsitz, *The Possessive Investment in Whiteness: How White People Profit from Identity Politics* (Philadelphia: Temple University Press, 2006), 107.
24. Thomas Sugrue, "Jim Crow's Last Stand: The Struggle to Integrate Levittown," in *Second Suburb: Levittown, Pennsylvania*, edited by Dianne Harris (Pittsburgh: University of Pittsburgh Press, 2010), 176.
25. Sugrue, "Jim Crow's Last Stand," 176.
26. Sugrue, "Jim Crow's Last Stand," 177.
27. Sugrue, "Jim Crow's Last Stand," 178.
28. Sugrue, "Jim Crow's Last Stand," 29.
29. Sugrue, "Jim Crow's Last Stand," 183.
30. Sugrue, "Jim Crow's Last Stand," 183.
31. Sugrue, "Jim Crow's Last Stand," 183–92.
32. Sugrue, "Jim Crow's Last Stand," 190.
33. Sugrue, "Jim Crow's Last Stand," 190.
34. Sugrue, "Jim Crow's Last Stand," 190.
35. Sugrue, "Jim Crow's Last Stand," 183.
36. Sugrue, "Jim Crow's Last Stand," 192.
37. Gil Scott-Heron, *Now and Then: The Poems of Gil Scott-Heron* (Edinburgh: Payback Press, 2000), 21.
38. Scott-Heron, *Now and Then*, 21.
39. Guy Debord, *Society of the Spectacle* (London: Rebel Press, 1983), 15.
40. Matthew Lassiter, *The Silent Majority: Suburban Politics in the Sunbelt South* (Princeton, NJ: Princeton University Press, 2006), 1.
41. Evan McKenzie, *Privatopia: Homeowners Associations and the Rise of Residential Private Government* (New Haven, CT: Yale University Press, 1994), 19.
42. Richard Heinberg, *The Party's Over: Oil, War, and the Fate of Industrial Societies* (Gabriola Island, BC: New Society Publishers, 2003), 7.
43. James Howard Kunstler, "Remarks," "PetroCollapse" Conference, New York, 5 October 2005, http://kunstler.com/other-stuff/petrocollapse-new-york-ny/.
44. Aimee Picci, "Most Americans Can't Handle a $500 Surprise Bill," *CBS Money Watch*, 6 January 2016, http://www.cbsnews.com/news/most-americans-cant-handle-a-500-surprise-bill/.
45. James Livingston "Fuck Work," Aeon, 2016, https://aeon.co/essays/what-if-jobs-are-not-the-solution-but-the-problem.
46. Huber, *Lifeblood*.
47. Elizabeth Kneebone, "Suburban Poverty Is Missing from the Conversation about America's Future," Brookings Institution, 15 September 2016. https://www.brookings.edu/articles/suburban-poverty-is-missing-from-the-conversation-about-americas-future/.
48. Richard Heinberg, *Powerdown: Options and Actions for a Post-Carbon World* (Gabriola Island, BC: New Society Publishers, 2008), 136–37.

49. Matthew Schneider-Mayerson, *Peak Oil: Apocalyptic Environmentalism and Libertarian Political Culture* (Chicago: University of Chicago Press, 2015).
50. Matthew T. Huber, "Too Much Oil," Jacobin, 22 March 2016, https://www.jacobinmag.com/2015/03/climate-change-fossil-fuels-oil/.
51. Erica Schoenberger, *Nature, Choice, and Social Power* (London: Routledge, 2015), 5.
52. Schoenberger, *Nature, Choice, and Social Power*, 5.
53. Andreas Malm, *Fossil Capital: The Rise of Steam Power and the Roots of Global Warming* (London: Verso, 2016).
54. Malm, *Fossil Capital*.
55. Karl Marx, *Capital*, vol. 1, translated by Ben Fowkes (London: Penguin, 1976 [1867]), 739.

CHAPTER 19

Vortex of Light (Ice Memoriam)

Maya Weeks

go north, they say
go north
pull the coal out of the ground
oil
fossils
under the midnight sun you can never get lost

*

mountains are scary
men are scary
gun violence
queer weather
how to be alive in 2016

*

a river of ice always moving
surge glacier unique to svalbard
routines unique to every glacier
krill in the water
filamentous plastic particles of different colors and types

*

digital checkpoints
burst pulses
pulsating glacier

*

amnesia in birds
seizures in sea lions
a crippling decline in the west coast shellfish industry

*

we drink poison every day

*

i don't want to pay to work
but i don't want to rot away

*

we lost years among sargassum
legs and boats covered in oil

*

paddling through kelp forest
across the ocean, which is exceptionally salty today

*

it's like we are landing on an island that is all the time getting smaller

*

BUT WHAT DO YOU LOVE?

*

enough garbage to fill the empire state building twice flowing into the ocean every week

*

We lost years among sargassum. Legs and boats covered in oil. The smaller the piece of plastic, the more available it is as food to animals. Plastic acts as a vector for other contaminants, such as heavy metals, such as the mercury released when coal is burned.

our concern is situated on the chemical level

moments of inertia and static

equilibrium

*

up here i don't eat liver from the animals i hunt, just to be on the safe side

*

adrift in the swell of international waters, both
absolutely into it and completely petrified
i want to crawl onto an ice floe and dive off it into the sea like a seal
the sun turning into a sheet of oil on the ice

*

drifting plastic

sea of jellies

eider duck nest made with plastic bag

*

warming oceans

increasingly acidic habitat

nothing to stop these nets from floating around

anything that gets trapped in it

the twenty people it takes to haul that fishing net up

once the reindeer get stuck

*

fish stocks that follow cooling waters

blue whales that follow fish into fjords

*

polar bears, their blood shot with endocrine disruptors, prevented from reproducing, unable to hunt without sea ice

*

Wonder how the plastic works as insulation for the ducklings. How about moisture? Does it affect survival?

*

if you go with a net
it's quite invisible
it's broken down into microplastic

*

more plastic than ice
more plastic than fish

*

80% FISHERY 20% OTHER

*

OUR MENTALITY IS THAT IT SHOULD BE QUICK AND SIMPLE; THAT NOTHING HAS VALUE EXCEPT OURSELVES

*

Everywhere we looked we found plastic.
The closer we looked, the more there was.
There were pieces of trawl everywhere.

Fish stomachs full of polyethylene.
Glaucous gulls with sores on the knuckles of their wings.
Kelp wrapped in transparent film.
A thin layer of gunk fraught with tinsel.

*

Even if you live a good life in relation to pollution, there are things you cannot see. A single washing of a fleece jacket can release thousands of tiny plastic particles. More than one hundred million microplastic particles are released into Advent Fjord every day.

*

colorful proof of our single-use world
the need for expediency above all else

CONTRIBUTORS

ACKROYD AND HARVEY are internationally acclaimed for time-based artworks that explore processes of organic and inorganic growth, change, and decay through a variety of media, including sculpture, photography, print, and film. Intersecting art with the disciplines of architecture, biology, ecology, and history, their work reveals an intrinsic bias toward process and event, reflecting scientific and environmental concerns that reference urban political ecologies, anthropogenic climate change, and biodiversity loss.

DARIN BARNEY is the Grierson Chair in Communication Studies at McGill University. He is the author and editor of several scholarly works including, most recently, *The Participatory Condition in the Digital Age* (University of Minnesota Press, 2016). Barney's current research focuses on materialist approaches to media and communication, infrastructure, and radical politics.

MARISSA LEE BENEDICT is a visual artist, writer, and independent curator/organizer. From the distillation of algal biodiesel to the extraction of a geologic core sample with a set of gardening tools, her visual art practice draws on traditions of American Land Art and systems aesthetics to critically observe social and material conditions of infrastructural space.

THOMAS BUTLER is a composer based in Glasgow, Scotland. His work uses found and archival sound, video, and instrumental performance to explore themes including environmental sustainability, psychogeography, and technology. Butler is musical director of new-music group Ensemble Thing and teaches musical composition at the University of Edinburgh.

JEFF DIAMANTI teaches literary and cultural analysis at the University of Amsterdam, and is the coeditor of *Materialism and the Critique of Energy* (MCM Prime Press, 2018), *The Bloomsbury Companion to Marx* (Bloomsbury Press, 2018), and the forthcoming *Climate Realism* (Routledge).

JACQUELENE DRINKALL is an artist, writer, independent curator, and lecturer in art at the University of Tasmania. Her art works with telepathy, energy transfer, immaterial labor, and socially/materially engaged extended cognition. Her writing is published in Artbrain.org, *Colloquy*, *Psychopathologies of Cognitive Capitalism*, *Leonardo Electronic Almanac*, and more.

KELLER EASTERLING is an architect, writer, and professor at Yale University. Her most recent book, *Extrastatecraft: The Power of Infrastructure Space* (Verso, 2014), examines global infrastructure networks as a medium of polity.

MEGAN GREEN is an artist from Newfoundland, who was part of a worker migration, associated with the oil sands, to Fort McMurray, Alberta, where she spent her formative years. She received a BFA from the University of Alberta and completed an MFA at the University of Waterloo in 2014.

MÉL HOGAN is an assistant professor of environmental media in the Communication, Media and Film Department at the University of Calgary. She has published on data centers and their environmental impacts in journals like *Television & New Media*, *First Monday*, and *Big Data & Society*.

CAMERON HU is a doctoral candidate in cultural anthropology at the University of Chicago.

MATTHEW HUBER is associate professor of geography at Syracuse University, where he studies political economy, historical geography, energy and capitalism, oil, resource governance, and social theory.

HANNAH IMLACH is a Scottish visual artist working predominantly in sculpture. Her practice is informed by current environmental research and often focuses on threatened ecologies and renewable energy transition. She has recently completed projects with the Royal Society for the Protection of Birds and the Not Just Energy Futures Research Group at the University of Edinburgh.

AM JOHAL is director of Simon Fraser University's Vancity Office of Community Engagement. He is the author of *Ecological Metapolitics: Badiou and the Anthropocene* (Atropos, 2015) and coauthor of *Global Warming and the Sweetness of Life: A Tar Sands Tale* (MIT Press, 2018).

JORDAN KINDER is an SSHRC doctoral fellow and PhD candidate in the Department of English and Film Studies at the University of Alberta.

ERNST LOGAR is an Austrian artist who has been working with photography, film, sculpture, and installations since 1995. In his work, he addresses existing power structures and historical, sociocultural, and ecological phenomena.

MARY ELIZABETH LUKA is assistant professor in arts, culture, and media at the University of Toronto. Her research interests include the modes and meanings of creativity in the digital age and investigations of how civic, artistic, scholarly, and business sectors are networked together.

JENNI MATCHETT currently studies critical conservation at the Harvard Graduate School of Design. Her design objectives challenge dominant energy paradigms via the intersection of renewables, architecture, the rural, and artistic practice. Previously, her professional experience was focused on sustainable commerce, solar energy development, and brand strategy.

CHRISTOPHER MALCOLM is a visiting faculty member in the environmental studies program at Humboldt State University. He completed his PhD in comparative literature at UC Irvine in 2017, with emphases in critical theory and visual studies. His current project is titled "Ecological Concessions: Environmental Damage and the Management of Harm."

MARIA MICHAILS is a Canadian artist working across science, technology, and the social sciences, creating projects that reimagine civic engagement with environmental issues. Her work has been exhibited and published internationally. She is currently a PhD candidate and doctoral fellow in electronic arts at Rensselaer Polytechnic Institute in Troy, New York.

ANTONIO NEGRI is an Italian Marxist sociologist and political philosopher who has taught philosophy and political science at the Universities of Padua and Paris. He has also been a political prisoner in Italy and a political refugee in France.

DAVID RUETER is an assistant professor at the University of Oregon in the art and technology program of the Department of Art. His creative practice makes use of new technologies, including custom software, custom electronics, data dumps and feeds, cartographic tools, and information systems, in media forms including sculpture, video, and performance.

IMRE SZEMAN is University Research Chair and Professor of Communications Arts, Geography and Environmental Management at the University of Waterloo.

DAVID THOMAS is a Joseph-Armand Bombardier Canada Graduate Scholar in the Department of English at Carleton University. In dialogue with sci-fi and speculative fiction, his dissertation analyzes the real-world impact of "social innovation" policy and explores how social scientists and humanists navigate the policy-made demise of academia's "two cultures."

MAYA WEEKS is a writer and artist from the rural central coast of California working on interactions between humans and the environment. Focuses include ocean ecosystems, climate change, gendered violence, and logistics.

JAYNE WILKINSON is managing editor of *Canadian Art*. A Toronto-based writer, editor, and curator, her work examines surveillance cultures, environmental politics, aesthetics, and representation in contemporary art practices. Previously, she was assistant curator at Blackwood Gallery (University of Toronto Mississauga) and editor/publisher of *Prefix Photo*.

INDEX